KOREA
SPECIALTY
COFFEE
GUIDE

KOREA SPECIALTY COFFEE GUIDE

First published 2022
by Tabi Books
6, World Cup-ro 28-gil, Mapo-gu, Seoul 03971, Korea
(E-mail: tabibooks@hotmail.com)

ISBN 979-11-92169-19-4 03590

Design : Lee Soo-joung
Print and Bind : Youngsinsa

* map source : National Geographic Information Institute
* Photos of Terarosa(Gangneung), Werk Roasters, Bean Brothers, Center Coffee, Cafe 304, Told A Story Coffee Roaster are provided by each company.
  The rest of the photos were taken by the authors.

| Charles Costello | Cho Won-jin | Shim Jae-beom |

KOREA
SPECIALTY COFFEE GUIDE

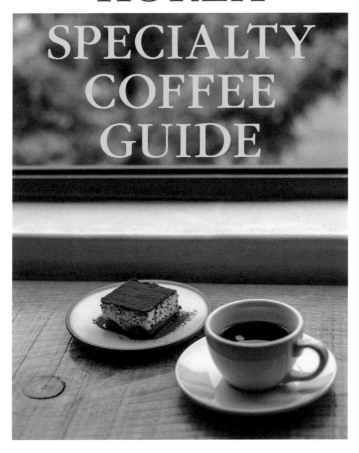

따비

Contents

Introduction

This is the first ever English language guidebook to South Korea's extraordinary specialty coffee industry. Here we present our 82 favourite cafes, expressing all the varied character of the country from Seoul, the sophisticated capital in the North-West, to Busan, bustling harbour city in the South-East and Jeju, tranquil island in the South.

After living and travelling in Melbourne, New York, London, and Tokyo, I was eager to explore the local coffee scene in Seoul while working here from 2019. Having learnt Korean studying five years previously in Gwangju, I teamed up with local writers Cho Won-jin and Shim Jae-beom and the project was born in August 2020. Originally starting with a list of some 200 cafes we filtered through as objectively as possible, completing the list of 82 cafes eventually in summer 2022.[*] These were the cafes that we thought captured the essence of the unique Korean coffee scene. We hope this guidebook will be useful to visitors, and Koreans alike, and that this collaboration is the first of many for us covering this fascinating country.

Charles

---

[*] The list of cafes was prepared in July 2022. Any changes to the cafes after the time of writing will be reviewed in future edits and subsequent versions of this book.

Timeline of Coffee in Korea

1886	First known record of coffee in Korea by American diplomat Percival Lowell (1855 - 1916), writer of the sketchbook and diary in Korea *Choson: Land of the Morning Calm* (《조용한 아침의 나라》).
1888	British vice-consul William Richard Carles (1848 - 1929) records having coffee with German Paul Georg von Möllendorff (1847 - 1901) in his anecdotal study *Life in Korea* (《조선풍물지》).
1897	Advert for "Java Coffee" in *The Independent* (《독립신문》) English language newspaper. First coffee to be drunk in Korea was almost certainly imported from Indonesia (later taken over by Brazil).
1899	Advert for one of the first cafes in the country, "Gyeong Shik-dang" (경식당) also in *The Independent*.
1900 - 1910	The king of Korea, Gojong (1852 - 1919) reportedly enjoyed coffee while entertaining foreign ambassadors at Deoksugung Palace (덕수궁).
1902	German-operated Sontag Hotel (손탁호텔), run by Marie Antoinette Sontag (1838 - 1922) serves coffee to foreign residents, diplomats and guests of Gojong of Korea.
1910 - 1945	Japanese occupation of Korea and the growth of cafe culture; venues like Cacadew (카카듀) open in 1927 and become popular with politicians, artists and students. Cafes from this era were focused predominantly around the Myeongdong (명동), Jongno (종로) and Chungmuro (충무로) areas of Seoul.
1950 - 1953	Introduction of instant coffee by American troops during the Korean War.
1959	Number of cafes or "Dabang" (다방) as they were colloquially known in Korea reaches 3,000.
1976	Dong Suh Foods Corporation (동서식품) launches their extremely popular instant coffee mix; made using pre-mixed sugar and cream this locally manufactured brand almost instantly overtook American coffee in popularity.

1999	Starbucks opens their first branch near Ewha Womans University (이화여자대학교).
2009 - 2016	Rapid increase in local coffee roasting companies and specialty coffee outlets led by brands Coffee Libre (커피 리브레), Namusairo (나무사이로), Fritz Coffee Company (프릳츠) and Felt (펠트). This period sees the beginning of the Korean "Third Wave" coffee boom.
2017	Founded in 2007, the annual World Barista Championship (WBC) is held in Korea for the first time.
2018	Hyundai Research Institute (현대경제연구원) reports that average coffee consumption in Korea is almost three times the global average, with the average person consuming 353 cups per year.
2019	WBC first Korean champion Jeon Joo-yeon (전주연) of Momos Coffee (모모스커피).
2021	20th Anniversary of the "Seoul Cafe Show" in COEX, now one of the largest coffee events in Asia (participated by 625 companies from 30 different countries and over 100,000 visitors).
2021	Starbucks opens their 1,500th cafe in Korea. The country is now the third largest cafe market in the world, after America and China.
2020-2021	Covid-19 puts Korea into lockdown with strict social distancing measures resulting in the forced closure of cafes. Record low sales of coffee particularly in the first lockdown in the Summer of 2020. Rise in "home cafe" culture marked by an increase in coffee subscription services, specialty coffee capsules and drip coffee bags.
2022	Local coffee market reaches record levels. There are now over 75,000 cafes across Korea, or the equivalent of one cafe per every 700 people.

Korea and the Rise of Specialty Coffee

The following is based on an excerpt from the book *Specialty Coffee, from San Francisco to Seongsu* (《스페셜티 커피, 샌프란시스코에서 성수까지》, 2022), by Cho Won-jin (조원진) and Shim Jae-beom (심재범).

Definition of Specialty Coffee and the "Third Wave"

"Third Wave" is a term invented by American futurist Alvin Toffler, to describe the character of society as it evolves through agricultural and industrial stages to a final post-industrial condition.

An attractive and succinct term, it quickly outgrew its original application and was first used in the context of coffee in 1999, by Timothy Castle, coffee broker and columnist for the Tea & Coffee Trade Journal Asia.

By 2008 coffee professionals including roasting pioneer Trish Rothgeb, green bean distributor Erna Knutsen, and Pulitzer Prize-winning food critic Jonathan Gold had developed the framework in three "Waves" as follows:

First Wave: Coffee becomes a household item.
Early coffee companies, including Folgers and Maxwell House, successfully market their products to the wider American public. With the invention and rise of popularity of instant coffee, first developed around the turn of the 20th century, coffee takes its place in every American household.

Second Wave: Rise of the American coffeehouse.
Coffeehouses including Peet's Coffee & Tea and Starbucks, from the 1970s,

expand their menu boards to show alongside traditional brewed coffee, a wide variety of espresso-based drinks, like cappuccinos, caffe lattes, and more recently, Frappuccino's.

Third Wave: Specialty Coffee and connoisseurship.
In San Francisco, in 1982, the American Specialty Coffee Association of America (SCAA) is established. By 1984 they produced a 100-point scoring system, referring to beans with 80 points or above as "specialty coffee". The unique characteristics of each bean, and the understanding of its origin becomes the primary focus of consumers or "coffee connoisseurs".

No longer just a simple commodity, the coffee experience now reflects the personality of the consumer. The role of baristas starts to change too, with a renewed focus on a wider variety of different brewing methods using fair trade and "specialty" coffee beans. With the gap steadily narrowing between farmer and consumer, the amount of information available to the coffee drinker becomes infinitely more traceable; from the coffee varietal to the producer who grew it, the altitude of growth and type of process used.

The Emergence of America's First Generation of Third Wave Cafes

The change in coffee consumer is behaviour was visible in popular culture too. Whereas in the 1980s popular sitcom "Cheers" was set in a Boston tavern, the comedy "Friends" of the 1990s revolves noticeably around the cafe "Central Perk".

The coffee industry in the 1990s grew rapidly as more and more people saw cafes as part of their living space, just like in sitcoms. Cafes such as Starbucks took the initiative in accepting them, leading the "Second Wave".

However standardised franchise cafes were inevitably insufficient to satisfy the tastes of every consumer. At the same time, consumers were starting to notice the difference between high street coffee houses and the cafes supplying specialty, higher quality and often "direct traded" coffee. A fan base quickly developed and the specialty coffee boom started to gather space. The first of these cafes are called "first-generation, Third Wave cafes", with representative examples including Chicago's Intelligentsia, Portland's Stumptown Coffee Roasters, and North Carolina's Counter Culture Coffee.

Intelligentsia Coffee & Tea founder Doug Zell, first saw the potential for growth in the coffee industry while working at San Francisco roastery Spinelli Coffee Company in 1989. At Spinelli, the centre of the second wave, he began a career as a barista, while raising capital for his own business. In 1995, Doug Zell and his partner Emily Mange moved to Chicago, opened the doors of "Intelligentsia Coffee & Tea" and gradually, almost single-handedly, elevated the standard of the city's coffee scene. Before long it became the number one roastery in the region, and later, a flagship brand for specialty coffee nationwide.

Unlike coffee from the second wave, characterised by a profoundly bitter, heavily roasted taste, Intelligentsia's coffee has a strong acidity associated with the original flavours and aromas of the coffee bean itself. Secondly, unlike the previous generation, which focused on espresso, fresh coffee is made using various brewing devices such as the Pour Over method, a variation of the hand drip method popular in Asia, including Chemex, V60, and Siphon. The Intelligentsia's sourcing team eventually started visiting coffee producing regions in the early 2000s, and after years of hard work, in 2003, began direct trading with a coffee farm in Huehuetenango, Guatemala. Twenty years later, Intelligentsia operates 15 different locations, runs a world class barista training lab and maintains relationships with over 50 different coffee farms across Central and South America, and Africa. Their original cafe is designed as a showroom, revealing Intelligentsia's brand identity, engaging with customers and showing their commitment to specialty coffee. The Monadnock Building branch, located inside a 19th century iconic Chicago skyscraper, reflects Intelligentsia's will to create a long-lasting sustainable coffee brand that co-exists with the local community.

When opening the Venice Beach store in Los Angeles, they designed a bar centred on the barista, drawing the focus on the story behind brewing coffee. Contrary to the customer-centric design of classic "Third Place" venues, the showroom concept reversed the tradition, putting coffee literally on centre stage and creating a theatre-like spectacle. This design concept is particularly prevalent in Korea, where brands play with minimalist interiors and unconventional designs to re-invent the social space of the cafe and excite social media with Instagrammable spaces.

Development of the Korean Specialty Coffee Industry

The history of specialty coffee in Korea began with three cafes, namely Terarosa (테라로사), and Namusairo (나무사이로) in 2002, and Coffee Libre (커피 리브레) in 2009. These three cafes not only introduced specialty coffee to the Korean coffee market, but also contributed to promoting Korean coffee culture to the world by interacting with experts on the global specialty coffee stage. In 2008, Terarosa Vice President Lee Yun-seon (이윤선) was appointed as the first judge of Cup of Excellence in Korea, and introduced various specialty coffees, including Ninety Plus Coffee, rare auction lots and micro-batch Geisha varietals. In 2013, Korea Brewers Cup Champion Jeong In-seong (정인성) achieved a runner-up result in the Sydney World Brewers Cup Championship using a Panama Geisha from Seoul-based brand Namusairo (나무사이로).

In 2008, Seo Pil-hoon (서필훈), an employee of Cafe Bohemian (카페 보헤미안, 1990 - 2021) in Anam-dong, became the first Korean to obtain Q-Grader accreditation, a coffee appraisal qualification issued by the Specialty Coffee Association of America (SCAA). A number of experts, Choi Jun-ho (최준호), quality control manager at SPC Group, and Kim Gil-jin (김길진), a Q-Grader judge, also started in the coffee business around this time. Seo Pil-hoon, after establishing Coffee Libre in 2009, was invited

by Yuko Itoi, the godmother of the Japanese specialty coffee industry, to become the second Cup of Excellence judge in Korea. Shortly after, he was joined by El Cafe Coffee Roasters (엘카페커피로스터스) and Momos Coffee (모모스커피), together building the foundation of specialty coffee in Korea.

Momos Coffee, opened near Oncheonjang Station (온천장역) in Busan (부산), has grown into one of the largest specialty brands not just in the Gyeongnam province, but across the country. Started by Q-grader certified Lee Hyun-ki (이현기), Momos Coffee inspired a coffee revolution in the region, influencing a wave of other brands to enter the scene; early front runners Blackup Coffee (블랙업커피) and In Earth Coffee (인얼스커피) were shortly followed by Werk Roasters (베르크로스터스) and Hytte Roastery (히떼 로스터리). Busan specialty coffee, led by Momos, spurred the development of regional coffee revolutions, including Coffee Place (커피플레이스) in Gyeongju (경주), Told a Story Coffee Roaster (톨드어스토리 커피로스터) in Daejeon (대전) and later on Coffee Temple (커피템플) in Jeju (제주).

After the success of the first generation of specialty coffee, a second generation of unique brands quickly followed including Hell Cafe Roasters (헬카페 로스터즈), Mesh Coffee (메쉬커피) and Naples-style espresso bar Leesar Coffee (리사르커피). Other cafes to mention in this post-2010 boom include 502 Coffee Roasters (502커피로스터스), Coffee Jumbbang (커피점빵, now Lowkey, 로우키), Millo Coffee Roasters (밀로커피 로스터스) and the aforementioned Coffee Temple (커피템플). Some of the signature menus created during this period, for example Temple's Yuja Americano (유자 아메리카노) or Millo's Mont Blanc (몽블랑) are still celebrated and enjoyed to this day.

Amid the continued growth of the Korean specialty coffee industry, companies with strong branding like Fritz Coffee Company (프릳츠) stood out from the crowd and quickly gathered a strong fan base amongst coffeeholics. Founded in 2014 by former Coffee Libre entrepreneur Kim Byeong-ki (김병기), and a team of five world class baristas, roasters and bakers, Fritz was an early hit on social media with it's retro cafe fit-out, cleverley branded merchandise and instantly recognisable seal mascot. The success of Fritz proved a great impetus for new establishments of local coffee shops across the country.

The increase in specialty coffee shops also provided an opportunity for the top brands to move into roasting coffee bean industry too; between them Fritz, Libre, Felt (펠트), Coffee Montage (커피몽타주) and 180 Coffee Roasters (180커피로스터스) started supplying roasted coffee beans to thousands of cafes in the burgeoning "Third Wave" coffee scene.

As Korean baristas started competing on the global brewing, roasting and barista circuit, they brought back with them a wealth of experience, adding even more depth to the market and creating industry-leading brands on their return. Lee Jong-hoon (이종훈) of Coffee Graffiti (커피 그래피티), Kim Sa-hong (김사홍) of Coffee Temple, and Jeon Joo-yeon (전주연) of Momos Coffee all became household names during this period, elevating the reputation of the Korean specialty coffee market on a global scale.

Korea is now a key market for major overseas coffee companies too. Opened in 2019 in Seoul , Blue Bottle (블루보틀) had a tremendous impact on the local coffee scene, forever changing the suburb Seongsu-dong (성수동) and putting a renewed spotlight on foreign coffee brands. The "Blue Bottle wave" was followed by a stream of Australian-backed brands including St. Ali, Market Lane and Dukes Coffee who attempted, albeit with varied success, to localise into the Korean market. Also from Australia, but established by Melbourne and Sydney-based Koreans respectively, ACoffee (에이커피) and Normcore Coffee (놈코어커피) recently established bases in Korea.

As of 2022, Korea is undoubtedly one of the most vibrant specialty coffee markets in the world. As the temporary pause on travel is lifted following Covid-19, we look forward to seeing the country garner more and more attention for its unique specialty coffee scene. Who knows, it might even be time for Korean brands to make a bid for overseas markets themselves, exporting another component in the growing and increasingly global K-content paradigm.

(Translated and edited for the *Korea Specialty Coffee Guide* by Charles Costello).

Cafe List (By Region)

20

Terarosa	테라로사	Daechi-dong	대치동	176
502 Coffee Roasters	502커피로스터스	Yeoksam-dong	역삼동	180
Leesar Coffee	리사르커피	Cheongdam-dong	청담동	182
Gray Gristmill	그레이 그리스트밀	Sinsa-dong	신사동	184

Incheon, Gyeonggi-do and Daejeon   page

Developing Room	디벨로핑룸	Cheonghak-dong (Incheon)	청학동 (인천)	190
Chromite Coffee	크로마이트커피	Ongnyeon-dong (Incheon)	옥련동 (인천)	192
Bean Brothers	빈브라더스	Dohwa-dong (Incheon)	도화동 (인천)	194
Cafe Tonn	카페톤	Chobu-ri (Yongin)	초부리 (용인)	198
180 Coffee Roasters	180커피로스터스	Yul-dong (Seongnam)	율동 (성남)	200
Coffee Nap Roasters	커피냅로스터스	Bongnam-ri (Pyeongtaek)	봉남리 (평택)	202
Told A Story Coffee Roaster	톨드어스토리 커피로스터	Galma-dong (Daejeon)	갈마동 (대전)	206

Busan and Gyeongsang Province   page

Werk Roasters	베르크로스터스	Jeonpo-dong	전포동	212
Treasures Coffee	트레져스커피	Jeonpo-dong	전포동	214
Hytte Roastery	히떼 로스터리	Jeonpo-dong	전포동	216
FM Coffee	에프엠커피	Jeonpo-dong	전포동	218
Blackup Coffee	블랙업커피	Bujeon-dong	부전동	220

Top 21 Cafes*

Cafe (English)	Cafe (Korean)	Region	Page
180 Coffee Roasters	180커피로스터스	Seongnam	200
502 Coffee Roasters	502커피로스터스	Seoul (Gangnam)	180
BlackUp Coffee	블랙업커피	Busan	220
Coffee Libre	커피 리브레	Seoul (North West, West)	140
Coffee Montage	커피몽타주	Seoul (North East, East)	114
Coffee Place	커피플레이스	Gyeongju	234
Coffee Temple	커피템플	Jeju Island	250
El Cafe Coffee Roasters	엘카페커피로스터스	Seoul (Central, North)	48
Felt	펠트	Seoul (Central, North)	44
FM Coffee	에프엠커피	Busan	218
Fritz	프릳츠	Seoul (North West, West)	164
Hell Cafe Roasters	헬카페 로스터즈	Seoul (Central, North)	50
Leesar Coffee	리사르커피	Seoul (Gangnam)	182
Mesh Coffee	메쉬커피	Seoul (North East, East)	106
Millo Coffee Roasters	밀로커피 로스터스	Seoul (North West, West)	130
Momos Coffee	모모스커피	Busan	228
Namusairo	나무사이로	Seoul (Central, North)	34
Onion	어니언	Seoul (Central, North)	82
Peer Coffee Roasters	피어커피로스터스	Seoul (North East, East)	102
Terarosa	테라로사	Seoul (Gangnam)	176
Werk Roasters	베르크로스터스	Busan	212

* For ease of reference the authors have organised the above Top 21 cafes in an ABC format.

Central,
North

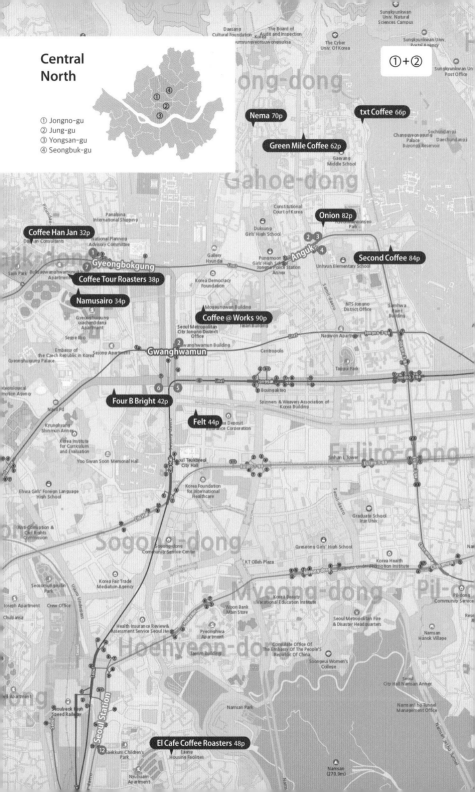

Central North

① Jongno-gu
② Jung-gu
③ Yongsan-gu
④ Seongbuk-gu

①+②

Nema 70p

txt Coffee 66p

Green Mile Coffee 62p

Onion 82p

Coffee Han Jan 32p

Second Coffee 84p

Gyeongbokgung

Coffee Tour Roasters 38p

Namusairo 34p

Coffee @ Works 90p

Gwanghwamun

Four B Bright 42p

Felt 44p

El Cafe Coffee Roasters 48p

Hakrim Dabang 72p

Hyewha

A Coffee 28p

Buam-dong

①

Sungshin Women's Univ.

Liike Coffee 76p

La Pluma & Bohemian 78p

Anam

Anam-dong

④

③

Bamaself 30p

Namyeong

Samgaki

Quartz 58p

Nakhasan Coffee 54p

Nothin Coffee 88p

Dongguk Univ.

Yongsan

Sinyongsan

Pont 56p

Travertine 60p

Itaewon

Hell Cafe Roasters 50p

ACoffee

에이커피

📍 19 Baekseok-dong 1ga-gil (Buam-dong), Jongno-gu, Seoul
🚇 N/A   🚌 Buam-dong Community Service Center   📷 @acoffee_seoul
🗓 **Mon to Fri** 10:00am - 6:00pm / **Sat to Sun** 11:00am - 7:00pm

Established **2020**   Brewing Method **Espresso, Filter (V60)**
Recommended Menu **Flat White**

Behind the Counter:
Machine **La Marzocco Linea 2 Group**   Grinder **Mazzer Kold S, Mahlkönig EK43**
Roaster **Probat Probatino, P12**

After five successful years in the heart of Melbourne's coffee capital Collingwood, Korean-owned ACoffee eventually opened a branch in Seoul in December 2020. Located in the suburb of Buam-Dong (부암동), surrounded by tranquil mountainside and art museums, this is the perfect place to escape the city for a much-needed flat white or pour over brew.

Designed by "Creative Studio Unravel", the interior is deliberately evocative of their Melbourne branch including minimalist seating, raw concrete walls, carefully curated flower arrangements and a bright white monochrome colour scheme.

Literally meaning "A Cup of Coffee", ACoffee was established in 2016 by barista Kang Byoung-woo (강병우) after a successful stint with local specialty coffee brands Saint Ali and Market Lane in Melbourne. Among his many achievements in the Australian

coffee industry, he is perhaps most famous for winning the notoriously challenging "World Cup Tasters Championship" in 2014.

The Seoul operation is run single-handedly by long term ACoffee business partner and seasoned barista Yoon Seong-jun (윤성준). Come on in for a chat, a perfectly made flat white and relax in the most Melbourne-like surroundings.

Coffee Beans The cafe doubles as a showroom for ACoffee's roasted beans which typically include a selection of four different single origins for both pour over and espresso.

"Melbourne in Seoul"

Bamaself

바마셀

📍 12 Wonhyo-ro 89-gil (Wonhyoro 1-ga), Yongsan-gu, Seoul   🚇 Namyeong (Line 1, Exit 1)
📷 @bamaself_coffee   🏬 **Mon, Tue, Fri** 10:00am - 7:00pm / **Wed** 10:00am - 3:00pm
Thur Close / **Sat to Sun** 11:00am - 7:00pm

Established **2020**   Brewing Method **Espresso**
Recommended Menu **Tricolore** (a unique coffee cocktail of Caffe Crema, Granita and Ice Cream)

Behind the Counter:
Machine **Faema E61 2 Group, MOAI Bar System One Group**
Grinder **Victoria Arduino Mythos Two**   Roaster **Diedrich IR-5**

"I only do espressos." Choi Hyeon-seon (최현선), owner and sole barista at this one-man coffee stand is a leading figure in the growing trend of Italian style espresso bars across Seoul. With its unique vibe and one of the most

diverse espresso menus in Korea, Bamaself (read: "By myself") has a cult following of baristas, specialty coffee lovers and roasters who come here for drinks including con panna (espresso with cream), con zucchero (with sugar) and Granita di caffe (iced espresso sherbet with double cream).

Bamaself is the culmination of over 13 years working in the industry and a flourishing barista competition career that includes a 1st Place at the 2010 Korean Barista Championship and 7th Place overall at the 2011 World Barista Championships (held in Bogota,

Venezuela). Beginning with Coffee Plant in 2008 Choi later went on to open the Five Extracts(파이브 익스트랙츠) franchise with business partner Do Hyeong-su(도형수), aptly named after the five most important flavour aspects of coffee: body, aroma, acidity, bitterness and sweetness.

After running the business for ten years Choi turned his attention to creating a new cafe legacy by establishing Bamaself, a tribute to the Italian espresso bars he first experienced on a trip to Turin, Milan, Verona and Padova in 2006. Opened in the summer of 2019 it became an instant neighbourhood favourite with locals gathering around the coffee service counter and outdoor bar area. The bright interior, with its green walls, yellow coffee bar and red Diedrich IR-5 roaster, are in stark contrast to the minimalist monochrome designs of Choi's former brand, 5extracts. With no takeaway available customers are encouraged to enjoy their drink by the bar in traditional Italian style. Get there early and you might be lucky enough to grab a croissant too, the perfect start to your day in Seoul.

Roasting in-house Choi carefully selects his specialty coffee from a range that includes Ethiopia, Guatemala,

Colombia, and Indian beans. Interestingly he uses two different espresso machines, a MOAI for single origin americanos and a classic Faema E61 for his espresso menus.

Local Tip The menu includes 14 different espresso-based drinks; just don't ask for an americano!

Coffee Han Jan

커피한잔

📍 16-1 Sajik-ro 9-gil (Pirun-dong), Jongno-gu, Seoul   🚇 Gyeongbokgung (Line 3, Exit 1)

📷 @sajikcoffeehanjan   📅 **Mon to Sat** 12:00pm - 9:00pm / **Sun** 12:00pm - 6:00pm

Established **2005**   Brewing Method **Espresso, Cold Brew, Filter** (V60, Siphon)
Recommended Menu **Hand Drip**

Behind the Counter:
Machine **San Remo**
Grinder **Ditting KR804, Mahlkönig Guatemala, Compak F10, Mazzer Robur**
Roaster **Custom**

Few cafes in Seoul come close to the history that Cafe Han Jan boasts. The vintage space is full of miscellaneous bric-a-brac from LP records, antique espresso cups and souvenirs from coffee growing regions around the world.

A world away from the digital grinders and automated espresso machines, Coffee Han Jan is one of the most analogue operated coffee spaces in the city. The counter boasts grinders that have all but disappeared from mainstream cafes including first generation Ditting KR804 and an original Mahlkönig Guatemala 710

commercial grade grinder.
Roasting in-house, owner Lee Hyeong-
choon (이형춘) stocks a wide variety
of coffee available from regions
including Ethiopia, Papua New
Guinea, Indonesia, Kenya, Nicaragua,
Costa Rica and Guatemala. After
selecting the region, you can choose
to have your coffee served by either
Hario filter, moka pot, Siphon or any

standard espresso menu.

Roaster Owner Lee uses a roaster
that originally started life at the
famous Hakrim Dabang (학림 다방).
After picking the item up in a spur-of-
the-moment garage sale he started
to roast part-time before selling
beans in the shop to regulars as early
as 2006.

Namusairo[*]

나무사이로

📍 21 Sajik-ro 8-gil (Naeja-dong), Jongno-gu, Seoul   🚇 Gyeongbokgung (Line 3, Exit 7)

📷 @namusairocoffee   📅 **Mon to Sun** 11:00am - 8:00pm

Established 2002 (Flagship branch opened here in 2013)
Brewing Method Espresso, Filter (V60), Cold Brew
Recommended Menu Pour Over, Hand Drip, Filter Coffee (Single Origin)
Okinawa Brown Sugar Cappuccino

Behind the Counter:
Machine La Marzocco GB5 3 Group   Grinder Mahlkönig EK43, Anfim SP II
Roaster Probat P25

One of the first specialty cafes in Seoul, Namusairo opened in 2002 before moving to its flagship location near Gwanghwamun (광화문) in 2013. Set in a renovated traditional hanok building, Namusairo's tranquil setting makes a stark contrast to the backdrop of towering skyscrapers from Seoul's nearby central business district. Namusairo has been making a name for itself in the coffee industry both domestically and overseas for over two decades now. They were famously the first Korean brand to reach the American market back in 2014 and received high praise particularly for their intriguing Love Letter blend. Today, customers come from all over Seoul to sample from their prized collection of rare single origins, small batch auction lots and signature drinks like the Okinawa Brown Sugar Cappuccino.

Namusairo currently roasts at their Bundang (분당) branch, south of Seoul on a trusty Probat P25 (Germany) roaster. When it comes to brewing the coffee, their Gwangwhamun branch has a formidable set-up on the counter; a La Marzocco GB5 espresso machine with Anfim SP II and Mahlkönig EK43 grinders.

Behind Story Namusairo in Korean (나무사이로) means "between the trees", a mention to the song "Path" by Kim Hwal-seong (김활성) which inspired owner and coffee expert Bae Jun-seon (배준선).

Interior Before Namusairo occupied this space in 2013, the building lay unoccupied for decades. Taking an experienced team of developers, the team managed to completely renovate the space in just three months, modernising the interior while protecting its heritage as much as possible. Note the internal courtyard, exposed wooden beams and giwa (tiled) roof.

Coffee Tour Roasters

커피투어 로스터스

📍 12 Sajik-ro 10-gil (Pirun-dong), Jongno-gu, Seoul　　🚇 Gyeongbokgung (Line 3, Exit 7)

📷 @coffee_tour　　🏠 **Mon to Fri** 8:00am - 9:00pm / **Sat to Sun** 10:00am - 8:00pm

Established　2008　　Brewing Method　Espresso, Filter (V60), Cold Brew

Recommended Menu　Single Origin Brewing Coffee / Nuvola (누볼라)

Behind the Counter:

Machine　Synesso MVP Hydra 2 Group

Grinder　Anfim SP II, Mazzer Robur S, Compak PK 100, Ditting KR804

Roaster　Probat Probatino, Probatone 5

Opened in 2008, Coffee Tour Roasters is the longest running specialty coffee brand in Seoul's central Jongno-gu

(종로구) district. Located just around the corner from Namusairo (나무사이로), the area has become somewhat of an inner-city cafe enclave with dozens of famous cafes now calling Gwanghwamun (광화문) home. From old-school drip coffee houses Coffee Tour Roasters and Coffee Han Jan (커피한잔) to trendy roastery Felt (펠트) and bagel cafe Four B Bright (포비 브라이트), there is something here to suit every coffee lover's taste buds. More and more local office workers, students and residents are gravitating to specialty coffee, urged on by brands like Coffee Tour Roasters

who now boast three cafes all within five minute's walk from the symbolic centre of Seoul, Gwanghwamun Square (광화문 광장).

Their largest branch, pictured here, doubles as their roastery too with a series of high-end Probat machines visible in the cafe's glass-walled laboratory. They typically have up to five single origins available at any one time, alongside three house blends and a decaf option. You can choose your espresso blend according to taste preference with each one roasted to a different strength level; Sound City (Strong), Take Five (Medium) and Sweet Spot (Light).

We recommend going for a filter coffee, but if you are in the mood for something sweeter, try Coffee Tour

Roasters' signature, the Nuvola (누
볼라), Italian for cloud. The base to
this creamy concoction also comes
in three different options including
Dutch, White (Iced Flat White) and Black
(Einspanner).

For non-coffee drinkers the cafe has
an extensive menu of herbal teas,
milk shakes and ades too. For a bite
to eat, take your pick from carrot
cake, homemade granola yoghurt
and freshly baked chocolate pecan
cookies.

Take a note of the stacks of burlap

sacks lining the walls, each stamped with the cafe's name and origin of each directly imported coffee bean. As well as being used in-store, most of Coffee Tour Roaster's prized imports are sold nationwide through the company's online distribution business.

Other Branches:

- **Coffee Tour Roasters (Gyeongbokgung - 경복궁)**
  - 📍 1 Jahamun-ro (Naeja-dong), Jongno-gu, Seoul
  - 🚇 Gyeongbokgung (Line 3, Exit 1)
- **Coffee Tour Roasters (Gyeonghuigung - 경희궁)**
  - 📍 12 Gyeonghuigung 2-gil (Naesu-dong), Jongno-gu, Seoul
  - 🚇 Gwangwhamun (Line 5, Exit 1)

Four B Bright

포비 브라이트

Lobby, Concordian Building, 76 Saemunan-ro (Sinmunno 1-ga), Jongno-gu, Seoul

Gwanghwamun (Line 5, Exit 6)   @fourb.hours

Mon to Fri 7:30am – 8:30pm / **Sat to Sun** 10:30am - 8:00pm

Established **2015**   Brewing Method **Espresso, Filter (V60)**   Recommended Menu **Flat White**

Behind the Counter:

Machine La Marzocco Linea 3 Group

Grinder La Marzocco Vulcano, Mahlkönig EK43S, Victoria Arduino Mythos One

Roaster Loring S15 Falcon, Giesen WP Series

Four B has been changing the daily coffee routine of countless office workers in Seoul since they first began in 2015 (using the Four B brand name since 2017). Located in spots that would usually be occupied by larger chain cafes like Starbucks, from skyscraper lobbies to apartment complexes, they now have 8 branches across the capital, as well as regional cities Paju (파주) and Cheongju (청주).

Influenced predominantly by Australian cafe culture, Four B's coffee menu has a classic mix of espresso-based drinks including Flat Whites, Cappuccinos and a Sydney favourite, the Piccolo Latte. Paired with 14 types of bagels (changing daily) and five different spreads, they have created a concept that is slowly breaking down the "Takeaway Iced Americano" cliché dominating most inner-city menus while playing an important role in the ever-changing landscape of Seoul's specialty cafe scene. As of 2020 Four B started roasting on a Loring S15 Falcon with a Giesen WP Series for smaller test batches. These include their signature blends Sweet Skunk and Smoker, as well as two to three rotational single origins at any one time. When it comes to pulling espresso shots, the counter set is a flawless line-up including La Marzocco Vulcano, Mahlkönig EK43S and Victoria Arduino Mythos One grinders. Espresso machines vary between branches with Four B Bright using a La Marzocco Linea in white stainless-steel finish.

Other Branches:

• **Four B (Gangnam - 강남)**

📍 106 Famille Station, Central City Shopping Mall, 205 Sapyeong-daero (Banpo-dong), Seocho-gu, Seoul

🚇 Express Bus Terminal (Line 3, 7 and 9, Exit 4)

• **Four B (Euljiro - 을지로)**

📍 100 Cheonggyecheon-ro (Supyo-dong), Jung-gu, Seoul

🚇 Euljiro 3-ga (Line 2, Exit 1)

• **Four B Basic (Hapjeong - 합정)**

📍 66 Yanghwa-ro 3-gil (Hapjeong-dong), Mapo-gu, Seoul

🚇 Hapjeong (Line 2 and 6, Exit 8)

"Basic, Best, Bright, Brilliant"

Felt *

펠트

📍 14 Cheonggyecheon-ro (Mugyo-dong), Jung-gu, Seoul

📷 Gwanghwamun (Line 5, Exit 5)   ⊙ @felt_seoul

🕐 **Mon to Fri** 8:00am - 9:00pm / **Sat to Sun** 10:00am - 9:00pm

Established **2015**   Brewing Method **Espresso, Filter (V60), Cold Brew**
Recommended Menu **Espresso (Classic)**

Behind the Counter:
Machine **Synesso MVP Hydra 3 Group**
Grinder **Mahlkönig E80 Supreme, K30, EK43**   Roaster **Probat P05, P25**

The minimalist theme that runs through all of Felt's cafes was born in their original Sinchon (신촌) branch (2015 - 2022) and continues with their most recent venture in the heart of Seoul's traditional business district, Jongno-gu (종로구). Located a short walk from Gwanghwamun station, the cafe sits directly opposite the Cheonggyecheon Stream (청계천), a 7 mile urban stream that runs through downtown Seoul.

Felt is distinctively discreet; the cafe is located between a convenience store, pharmacy and commercial office space, only noticeable by a small marble plaque on the outside of the building. A beech wood bench runs round the entire length of the cafe with alternative seating at the chrome topped central counter. The walls are covered with a dramatic red

stage curtain, designed to dampen the sound of the concrete space while adding to the stage-like atmosphere of this unique venue.

The trio behind Felt coffee, head roaster Kim Young-hyun (김영현), barista Song Dae-woong (송대웅) and Jeong Hwan-sik (정환식), originally started out with a brand called "Mad Coffee" in Seoul's Yeouido district. Run concurrently with an off-site roasting factory, the business soon began to take off and within a matter of months started supplying beans to a whole host of cafes in the burgeoning "third wave" coffee scene. Before long Song decided it was time to re-brand, expand the focus on specialty coffee

and deliver a brand-new coffee-drinking experience under the name, Felt. Famous for its unconventional design strategy, the brand has sent shockwaves through the industry ever since day one and continues to excite local cafe goers.

Felt sources all its coffee through direct trade and hand picks batches using their in-house quality control team of green bean buyers, cuppers, and roasters. Now in its seventh year the team now roast an impressive 70 tons annually and supply beans to 300 plus cafes across the country.

In-store, Felt serves up three different espresso varieties including Seasonal, Classic and Decaf, all prepared

individually using their single dosage K30 and Mahlkönig E80 grinders.

Audiophile Positioned either side of the coffee bar, carefully suspended from the ceiling, two Western Electric 30154 Horn speakers play out Felt's signature relaxing vibes.

Branding Courtesy of Lee Jae-min (이재민) at Studio FNT, the packaging was redesigned in 2018, resulting in their signature criss-crossed lines on white backgrounds.

Gwanghwamun Branch Considered their flagship branch, Felt Gwangwhamun opened in October 2018 and has since been almost single handedly responsible for changing the "morning coffee" routine of many commuters to this CBD area of the city. Opening at 7am, this is a great spot to pick up both espresso and drip coffees as well as baked goods and coffee beans for home brewing.

Other Branches:
- **Felt (Dosan-gongwon** - 도산공원**)**
    - 📍 23 Eonju-ro 164-gil (Sinsa-dong), Gangnam-gu, Seoul
    - 🚇 Apgujeong Rodeo (Suin-Bundang Line, Exit 5)
- **Felt (Gwanghwamun** - 광화문**)**
    - 📍 Basement 2, Gwanghwamun D Tower, 7 Jong-ro 3-gil (Cheongjin-dong), Jongno-gu, Seoul
    - 🚇 Gwanghwamun (Line 5, Exit 3)

Felt (Gwanghwamun)

El Cafe Coffee Roasters*

엘카페커피로스터스

📍 68 Huam-ro (Huam-dong), Yongsan-gu, Seoul
🚉 Seoul Station (Line 1 and 4, Airport Line, Gyeongui-Jungang Line, Exit 12)
📷 @elcafecoffeeroasters
🗓 **Wed** 9:00am - 5:00pm / **Thur to Fri** 11:00am - 5:00pm / **Sat to Sun** 11:00am - 7:00pm

Established 2010   Brewing Method Espresso, Filter (V60, Poursteady), Cold Brew
Recommended Menu Classic Blend Americano

Behind the Counter:
Machine MOAI Bar System 3 Group
Grinder Anfim SP II, Eureka Mignon Specialita, Mahlkönig EK43   Roaster Probat

A trend-setting roastery in the Korean specialty coffee industry, El Cafe was established in 2010 by seasoned professional Yang Jin-ho (양진호). Established around the same time as Coffee Libre, El Cafe was one of few companies at the time travelling to

coffee growing regions and sourcing green beans direct from source. Up until the outbreak of Covid-19, Yang would spend between two and three months of every year on the road, inspecting crops, building relationships with the farmers and cupping lots during auction season. As a result, El Cafe is an excellent place to try rare auction lots and premium Geisha single origins. Recent imports include unique micro batches like Nicaragua "Jesus Mountain", Panama "Savage Anthem" Geisha, and Colombia Bella Vista. Aside from their top-shelf range, El Cafe's brew

bar stocks up to 15 different types of single origins and house blends ("Classic" and "Italian Job"). After seven years (2015 to 2022) running the cafe from their HQ and roastery warehouse in Seoul's Yangpyeong-dong (양평동), the brand recently relocated over the Han River to start a new chapter in Huam-dong (후암동). For the quickest route there, head out of exit 11 of Seoul Station (Lines 1 and 4, Airport Line, Gyongui-Jungang Line), opposite Seoul City Tower, and walk up the hill in the direction of Namsan Tower on the busy thoroughfare, Huam-ro (후암로).

Home Barista Market One of the more successful examples of online coffee retailers in Korea, El Cafe runs a comprehensive e-commerce site selling everything you need to set up the perfect "home cafe". Yang regularly uploads brewing lessons, roasting tips and short seminars to the cafe's YouTube channel "elcafecoffee" (엘카페 양사장). If you are shopping in-store, all their coffees are available to purchase in 200 and 500g bags; simply ask the team for advice and the baristas will happily provide tips on grind size, ratios and brewing method.

Hell Cafe Roasters*

헬카페 로스터즈

📍 76 Bogwang-ro (Bogwang-dong), Yongsan-gu, Seoul
🚇 Itaewon (Line 6, Exit 3)   📷 @hellcafe_roasters
📅 **Mon to Fri** 8:00am - 10:00pm / **Sat to Sun** 12:00pm - 10:00pm

Established 2013   Brewing Method **Espresso, Filter** (Nel Drip Pot)
Recommended Menu **Hell Latte**

Behind the Counter:
Machine **Slayer Espresso 2 Group**
Grinder **Mazzer Robur**   Roaster **Probat P05, P12**

Founding members Lim Seong-eun (임성은) and Kwon Yo-seob (권요섭) joined forces in the late 2000s, after meeting at the first anniversary party of Coffee Libre in Yeonnam-dong. In time, they both closed their existing cafes, combined their barista and roasting skills, and created the Hell Cafe brand in 2013. Now almost ten years old the cafe has a massive following as the go-to place for Itaewon's coffee aficionados.
As well as the popular Hell Latte, many regulars come here to sample Kwon's nel drip coffee, made using their classic dark roast house blend.

Prepared like a typical pour over, the nel drip uses a cloth filter and is prepared with a higher coffee-to-water ratio to produce an aromatic coffee with an almost espresso-like texture.

Retro Touch Don't expect Spotify to run the playlist here, Hell Cafe's music comes from a huge library of CDs and LPs, played out on vintage surround sound systems.

Flower Arrangements In another

unique analogue touch, the flower arrangements for each table are changed every week.

Roasting Brand Originally a micro-roastery, roasting everything by hand, the team recently upscaled their presence in the B2C market with a nationwide distribution business and a range of "home-cafe" goods for online shopping including capsule coffee and drip coffee sets.

Hell Cafe Spiritus The cafe is located on Ichon-dong's main commercial strip in a lowkey suburban two story 1970s building. The interior concept is simple, characterised by deep

mahogany furniture, a long cocktail bar and heavy blinds blocking out the streetlights. Huge Tannoy and Klipsch speakers play baroque classical music, while award-winning baristas add the finishing touches to your latte directly at your table.

Other Branches:
• **Hell Cafe Bottega**
  📍 33 Wonhyo-ro 89-gil (Wonhyoro 1-ga), Yongsan-gu, Seoul
  🚇 Namyeong (Line 1, Exit 1)
• **Hell Cafe Spiritus**
  📍 31-208 Han River Mansion Condominium Complex, 248 Ichon-ro (Ichon-dong), Yongsan-gu, Seoul
  🚇 Ichon (Line 4, Exit 3-1)

Nakhasan Coffee

낙하산커피

📍 25 Hangang-daero 46-gil (Hangangno 2-ga), Yongsan-gu, Seoul
🚇 Samgakji (Line 4 and 6, Exit 3) 📷 @nakhasan.coffee
🕐 **Mon to Sun** 11:00am - 7:00pm

Established **2020**   Brewing Method **Espresso, Cold Brew**
Recommended Menu **Nakhasan Latte (Iced Americano Slushie)**

Behind the Counter:
Machine **Nuova Simonelli Appia Life 2 Group**
Grinder **Mazzer Robur**

Barista and owner Lee Hoon (이훈) has worked in the cafe industry for over 15 years now, with experience at a whole host of different cafes from B.A.O.K (Barista Association of Korea), popular chain coffee shop Paul Bassett and the highly influential Hell Cafe. Settling on Yongnidan-gil (용리단길) for his first venture Nakhasan, the cafe has a huge following for its creative menu of iced coffees and signature slushies.

Close to Sinyongsan (신용산) and Samgakji (삼각지) stations as well as Seoul's KTX bullet train hub Yongsan (용산) Station, the area has traditionally been somewhat of a transitory area. Over the last few years, however, the central suburb has quickly gentrified with a growing number of trendy restaurants, wine bars and cafes now occupying the neighbourhood's alleyways.

The menu here has over 30 different items including one of the largest

selections of iced coffee varieties in Seoul. From Dutch coffee to shakeratos, iced flat whites and chilled americanos, the skilled Nakhasan barista team has got you covered. The usual fare of cappuccinos, lattes and espressos are also available, made using their own blend supplied by long-term friend and business partner Second Coffee (세컨드커피). For non-coffee drinkers Nakhasan also sells refreshing milk slushies, ice chocolate and a range of ice teas. Whatever you order to drink, don't miss out on their epic home-baked clotted cream scones and butter fudge.

Retro Vibes A fine example of Seoul's booming independent cafe market, Nakhasan Coffee is full of unique finishing touches, signature menus and on-point retro branding.

Iced Americano Culture Ever since coffee became popularised in the early 2000s, the go-to order for the majority of cafe-goers has always been the Iced Americano (shortened to 아아 or "ah-ah" in Korean). Enjoyed all year-round Koreans prefer this simple menu as a quick, grab and go caffeine solution that both quenches your thirst and saves time on the weekday coffee run.

Pont

폰트

📍 19-16 Hangang-daero 15-gil (Hangangno 3-ga), Yongsan-gu, Seoul
🚈 Yongsan (Line 1, Gyeongui-Jungang Line, Exit 1) or Sinyongsan (Line 4, Exit 3)
📷 @pont_official_   📅 **Mon to Sun** 11:00am - 9:00pm

Established **2020**   Brewing Method **Espresso, Filter (V60)**
Recommended Menu **Single Origin Brewing Coffee**

Behind the Counter:
Machine **Synesso MVP Hydra 3 Group**
Grinder **Mahlkönig EK43, Anfim SP II**   Roaster **Diedrich IR-5**

Opened in July 2020, Pont's flagship branch cleverly fits between two alleyways in Yongsan's main market

street Hangang-daero (한강대로). Originally a railway office, built in the 1940s, the large 105m² space was transformed by Seoul-based interior design firm "Studio Stof" (Park Seong-jae), the same brand behind recent cafe renovations including Felt (펠트) (Gwanghwamun), Cerulean (세루리안) (Insadong) and 502 Coffee Roasters (502커피로스터스) (Gangnam).

Brightening up the space with creative alcove lighting and relaxing shades of cream and terracotta throughout, the cafe maintains a whole host of original features from rustic brick facade, exposed timber trusses and

steel support beams. Stretching from the front door all the way into the middle of the cafe, a long beech wood counter bridges the space between barista and customer, while representing Pont's role in bridging the gap between coffee producer and consumer by serving only direct trade beans.

Pont serves two espresso house blends Overtime and Hoist, with at least four specialty graded single origins available for filter at their signature pour-over bar.

Located just a short walk from Seoul's major railway station, Yongsan (a high-speed KTX hub), this is an ideal place to introduce yourself to the capital's coffee scene. Other options nearby to consider include Travertine (트래버틴), Quartz (쿼츠) and Nakhasan Coffee (낙하산커피).

Other Branches:
- **Pont (Mullae - 문래)**
  - 6 Gyeongin-ro 77ga-gil (Mullae-dong 2-ga), Yeongdeungpo-gu, Seoul
  - Sindorim (Line 2 and 1, Exit 6)

Quartz

퀴츠

📍 158 Hangang-daero (Hangangno 1-ga), Yongsan-gu, Seoul
🚇 Samgakji (Line 4 and 6, Exit 3)   📷 @_____quartz
🗓 **Mon to Sun** 11:00am - 7:00pm

Established **2016**   Brewing Method **Espresso, Filter (V60)**
Recommended Menu **Quartz Latte / Quartz Mocha / Quartz Caramel Macchiato**

Behind the Counter:
Machine **La Marzocco Linea 3 Group**
Grinder **Anfim SP II**   Roaster **Giesen W6**

Originally opened as a roastery in Seongbuk-gu (성북구), owner Yoo Yeon-joo (유연주) relocated the business to its current showroom in Yongsan-gu (용산

구) in 2020.

Before first setting up shop in 2016, Yoo became the first ever female national barista champion by winning gold medals respectively in the 2012 Korea National Barista Championship (KNBC) and the 2015 Korea Brewers Cup (KBC). Following Yoo's success the brand quickly started to attract the attention of coffee enthusiasts and queues soon started to appear outside the cafe's hip minimalist lounge. Rather than focusing on hand drip coffee, Quartz's menu is based predominantly on classic espresso variations as well as teas, ades and hot

chocolate for the non-coffee drinkers. We recommend going for one of the trio of delicious signature menus; prepared using their own zero-additive condensed milk, Quartz's trademark caramel macchiato, mocha and latte have a distinctive sweet taste and soft creamy texture. When it comes to the coffee beans, Quartz uses a combination of Colombian, Ethiopian, Indian and Costa Rican to create their core range of medium-dark blends Nostalgia, Quartz and Old Fashioned.

Alongside local favourites Pont (폰트), Nakhasan Coffee (낙하산커피) and Travertine (트래버틴), this is another great option to add to any tour of Yongsan's popular Yongnidan-gil (용리단길) cafe street.

Travertine

트래버틴

📍 18-7 Hangang-daero 7-gil (Hangangno 3-ga), Yongsan-gu, Seoul

🚇 Yongsan (Line 1, Gyeongui-Jungang Line, Exit 1), Sinyongsan (Line 4, Exit 3)

📷 @travertine_cafe　🕐 **Mon to Fri** 12:00pm – 8:30pm / **Sat to Sun** 11:00am - 9:00pm

Established **2018**　Brewing Method **Espresso, Filter (V60)**
Recommended Menu **Filter Brew**

Behind the Counter:
Machine **Synesso S200 2 Group**　Grinder **Mahlkönig EK43, Anfim SP II**

Popular local hangout on Yongsan-gu's trendy Hangang-daero (한강대로), Travertine has been at the heart of this community for close to four years now. Built inside an existing hanok (traditional residential building), Travertine's interior maintains several original features including 100-year-old wooden roof beams, tiled giwa rooftop and red brick walls. Once through the main porch, you'll find the cafe raised from the ground in a glass box structure, surrounded by a gravel moat and quirky alcove seating areas. The inside features two open planned spaces, both focused on large communal seating areas, an open service counter and shelving full of goods and coffee from partner roaster La Cabra (Denmark).

Travertine has worked with La Cabra

60

since day one, favouring the world-famous Copenhagen-based roaster for its light North European style. Choose from a selection of up to five carefully selected La Cabra single origins. The cafe has added a complete series of Costa Rican beans to their 2022 line-up, including a Geisha (Tarrazu, Honey Process), micro-lot Washed (West Valley, Zuniga Brothers) and a Natural (Tarrazu)

from Cup of Excellence award winning farm "Divino Nino".

Heart of the community Placing an emphasis on engaging with the community, owner Lee Seung-mok (이승목) deliberately created communal seating spaces within the cafe including long corner sofas and a central low-level stone bench.

Green Mile Coffee

그린마일 커피

📍 64 Bukchon-ro (Gahoe-dong), Jongno-gu, Seoul
🚇 Anguk (Line 3, Exit 2)   📷 @green_mile_coffee
🕐 **Mon to Fri** 8:00am - 7:00pm / **Sat** 10:00am - 8:00pm / **Sun** 10:00am - 7:00pm

Established 2009   Brewing Method Espresso, Filter (V60, Siphon)
Recommended Menu Siphon Coffee / Green Mile Latte (Matcha + House Blend)

Behind the Counter:
Machine MOAI Bar System 2 Group
Grinder Mahlkönig E80, Mazzer Robur   Roaster Giesen W15

First generation coffee roasting brand, Green Mile Coffee was established in Seoul's Nonhyeon-dong (논현동) in 2009. After ten years servicing the Gangnam area, the team decided to close the original branch and relocate north of the Han river in the picturesque Bukchon Hanok village (북

촌한옥마을).

Located at the northern end of the village, the charming hilltop surroundings are steeped in 600 year old history from Gyeongbokgung Palace (경복궁), Changdeokgung Palace (창덕궁) and the Jongmyo Royal Shrine (종묘) all nearby. Head to the rooftop of the cafe for a view of this historic urban environment with endless hanok house rooftops and winding alleyways below. Get your bearings of the city as you gaze out further across the sprawling Seoul cityscape; on a clear day you can even see as far as the Namsan Tower! When it comes to the coffee, siphon brewers take centre stage at both the Bukcheon and Jeju venues. Originally invented in Europe in 1830, the siphon brewing technique uses a dramatic

boiling process to draw boiling water through a glass chamber into a vacuum thereby fully immersing the coffee and creating a clean, full-bodied cup.

For siphon you have the choice of up to four different single origins, while espresso-based drinks are served using a trio of in-house roasted blends, Song Bird, Mono Gram and Old School.

Traditional touches Note the waiting buzzer, designed in the same style as a Chosun dynasty wooden identification tag (호패 모양).

Other Branches:

• **Green Mile Coffee (Aewol - 애월)**

137-1 Aewol-ro (Aewol-ri), Aewol-eup, Jeju-si, Jeju-do

N/A

txt Coffee

txt 커피

📍 121 Changdeokgung-gil (Wonseo-dong), Jongno-gu, Seoul
🚇 Anguk (Line 3, Exit 3)　　📷 @txtcoffee
🕐 **Tue to Sat** 11:00am - 5:00pm / **Sun, Mon** Closed

Established **2017**　　Brewing Method **Espresso, Filter (V60)**
Recommended Menu **Brewing Coffee**

Behind the Counter:
Machine　**Slayer Espresso 1 Group**
Grinder　**Mahlkönig EK43, Victoria Arduino Mythos One**　　Roaster　**Giesen W1**

Set deep in the tranquil backstreets of Bukchon village (북촌마을), txt Coffee ('txt') has quietly been serving up premium single origin filter coffee and flat whites since 2017. Located just a stone's throw away from the grounds of UNESCO

World Heritage site Changdeokgung palace (창덕궁), this is a must-visit cafe for anyone looking to soak up some culture with their morning brew. Masterminds of site-integrated architecture, Seoul-based "The First Penguin" have taken care to preserve the heritage of this important neighbourhood with 'txt'. In fact, it blends in so seamlessly with the surroundings, you could easily stroll past the unassuming exterior, but don't! Behind the counter, Lee Soo-hwan (이수환) runs an impressive one-man operation roasting micro-lots on a classic Giesen W1, before grinding, brewing and serving each order

individually. You can either opt for an espresso-based white coffee or choose from a selection of premium single origin filter options, noting your taste preferences on their detailed order form* using the pencils provided. Try and pinpoint your ideal level of sweetness, bitterness or the type of crema and hand over the counter for a personalised caffeine prescription.

* The form comes in Korean, but also in English and Japanese.

Buying directly from highly competitive auctions Best of Panama, Cup of Excellence and Esmeralda Special Auction, 'txt' often boasts some of the highest SCA (Specialty Coffee Association) scoring coffees of any cafe in Seoul. Past favourites include Panama Elida Geisha ASD Natural, Costa Rica Don Cayito and Panama Esmeralda Noria Micro-Lot. If you prefer espressos, you are in for a treat as well. Lee has a finely tuned

single group Slayer espresso machine and serves all his coffees ristretto-based as opposed to the standard single and double espresso recipes. This is purely the preference of the barista who has finely tuned his single group Slayer espresso machine to suit the finer grind and slower extraction required for shorter espresso shots (simply put, Ristretto means a "restricted"

shot of espresso).

Interior The interior space focuses on Korean touches with Koryo celadon green cups, vintage medicine shop dispensary shelving (한약방), and uninterrupted views of hanok rooftops and Korean pines through floor-to-ceiling windows.

Nema

네마

📍 86-4 Samcheong-ro (Samcheong-dong), Jongno-gu, Seoul
🚇 Anguk (Line 3, Exit 2)   📷 @nemacoffee
📅 **Mon to Fri** 8:00am - 8:00pm / **Sat to Sun** 10:00am - 8:00pm

Established **2021**   Brewing Method **Espresso, Filter (V60)**
Recommended Menu **Single Origin Flat White**

Behind the Counter:
Machine **Slayer Espresso 2 Group**
Grinder **Mahlkönig EK43, Victoria Arduino Mythos One**

Ten minutes' walk from Seoul's central business district Jongno-gu (종로구), the northern suburbs of this metropolis are full of old-world charm. Spread across Samcheong-dong (삼청동), Ikseon-dong (익선동) and Sajik-dong (사직동), these traditional areas feature centuries-old hanok alleyways and endless waves of curving giwa roofs

reaching out above the low-lying skyline. After years of neglect these historic neighbourhoods are going through a renaissance with younger generations moving in, creating cafes, shops, and restaurants in carefully renovated Korean houses.

Located just a few steps off the main Samcheong-ro street (삼청로), Nema is a perfect example of how cafe entrepreneurs are revitalising these areas, combining traditional Korean architecture with contemporary specialty coffee culture. Maintaining the original appeal of the building, the design team here have stripped

back this space to its original structure, complete with exposed timber beams, internal courtyard and sloping tiled roof. Wall-length windows on both street facing sides of the building add a modern touch, providing postcard views of Samcheong-dong and flooding the space with natural light.

Opened in 2021, Nema is run by former Fritz (프릳츠) Coffee Company baristas Kim Sa-gon (김사곤) and Chae Su-hwi (채수휘). Building on a decade-long career in the industry, owner Chae serves up a classic menu of espresso-based drinks and single origin filter brews using specialty beans from Chromite Coffee and other partner brands. Nema isn't the only cafe run by ex-Fritz members; other successful "second generation" cafes include Reifen Coffee (라이픈 커피) in Sokcho (속초), Chowdy (차우디) in Ilsan (일산), as well as Pont (폰트) and Bake Basic (베이크베이직) in Seoul. Alongside brands like Coffee Libre (커피 리브레), Felt (펠트) and Namusairo (나무사이로), Fritz is playing an immense role in the rise of "third wave" coffee culture, both directly and indirectly through this steady flow of veterans entering the independent cafe scene.

Hakrim Dabang

학림 다방

119 Daehak-ro (Myeongnyun 4-ga), Jongno-gu, Seoul
Hyewha (Line 4, Exit 3)   @hakrim_coffee
Mon to Sun 10:00am - 11:00pm

Established 1956   Brewing Method Espresso, Filter (Chemex), Cold Brew
Recommended Menu Royal Blend Coffee

Behind the Counter:
Machine La Cimbali M39 GT Dosatron 3 Group
Grinder Mahlkönig K30   Roaster Probat L5

Like Rome's Antico Caffe Greco, or Paris' Café de Flore, Seoul's longest running cafe Hakrim Dabang has played a significant role in the context of South Korea's modern history. Established in 1956 the cafe is now in its seventh decade and is recognised by the local government as an integral part of "Seoul Future Heritage" (서울미래유산) due to its role in the democratisation movement that took place in the 1980s. Located close to the original Seoul National University campus the cafe was a favourite for poets, writers and later student activists who used the cafe for their National Democratic Students League (전민학련) meetings.

Running the cafe since 1987, the fourth owner of Hakrim Lee Chung-ryul (이충렬) is a true veteran of the local coffee scene. Like many of the first generation Korean coffee roasters, Lee learnt his trade initially in Japan before returning to Seoul where he effectively revolutionised the city's coffee drinking habits.

At a time when instant coffee was peaking in popularity, he started roasting "Hakrim Blend" coffee beans on-site, introduced new filter coffee menus and began distributing whole beans to friends, apprentices and customers across Seoul. He set up a classroom too, educating colleagues on all aspects of coffee craftsmanship from roasting techniques, importing green beans and the art of hand drip brewing.

In recent years the cafe has appeared in a number of K-Dramas (including "A Love From Another Star", 별에서 온 그대) and films, popularising the location with tourists and younger Korean visitors alike. The interior aesthetic and atmosphere remains much the same as it was in the 1950s and 1960s;

tiered mezzanine seating, vintage LP player, graffiti covering the walls and dark wooden booths full of students enjoying house blend drip coffee and cream cheese cakes. This is a must-visit spot for anyone looking to add some old-world charm to your Seoul cafe crawl.

Coffee and Cheese cake This classic combination is a must-order treat; choose from one of four types of blended coffee including Americano,

Regular, Strong and Royal Blend. We recommend going for a "Royal Blend" paired with the Blueberry Cheese cake.

Interior Opened in 1956, much of the interior of Hakrim Dabang is original, dating back to the bygone era of the city's "dabangs", or traditional tea and coffee houses. The quirky mezzanine floor features old-school booth tables, cosy alcove seating and wooden desks scrawled with decades old graffiti.

Liike Coffee

리이케 커피

📍 24 Bomun-ro 34ga-gil (Dongseon-dong 3-ga), Seongbuk-gu, Seoul
🚇 Sungshin Women's University (Line 4, Exit 1)   📷 @liike_coffee
🏪 **Wed to Mon** 12:00pm - 6:00pm / **Tue** Closed

Established **2018**   Brewing Method **Filter (V60)**
Recommended Menu **Drip Coffee**

Behind the Counter:
Machine **N/A**   Grinder **Mahlkönig EK43**
Roaster **Probat P05**

Quiet neighbourhood roastery and showroom, Liike is somewhat of a local landmark for coffee lovers in the university neighbourhood of Dongseon-dong (동선동). Lee Yoon-haeng (이윤행) and Ha Ji-yeon (하지연), the coffee duo have perfected the open-plan space to carefully create

a homely living room cafe vibe. With no espresso machine to disrupt the peace and quiet, Lee and Ha serve only hand drip filter coffee alongside two classic menus, cafe au lait and affogato.

Roasting in the back room, Lee has up to five different single origin beans at any one time, many of them coming directly from premier auction houses in Hawaii (ISLA Custom Coffees Private Collection), Yemen (QIMA Specialty Coffee) and Costa Rica (Cup of Excellence).

Tasting Notes No need to decode complicated coffee notes here,

each coffee comes instead with a corresponding coloured paper and short descriptive summary.

Scandinavian Vibes The interior is fitted out in a subtle array of cream and yellow colours, interspersed with artisan Danish furniture and finished off with bright floor-to-ceiling windows. In fact, the the name itself, Liike (리이케 in Korean), is Finnish for shop.

Menu Stripped back to a simple trio of classics, choose from either hand brew single origin drip coffee, cafe au lait or affogato.

La Pluma & Bohemian

라플루마앤보헤미안

📍 18 Goryeodae-ro 27-gil (Anam-dong 5-ga), Seongbuk-gu, Seoul 　　🚇 Anam (Line 6, Exit 2)

📷 @bohemian_anam 　　🗓 **Mon to Fri** 12:00pm - 5:00pm

Established　1990 / Re-opened 2021 as La Pluma
Brewing Method　Espresso, Cold Brew, Filter (V60)
Recommended Menu　Dutch Coffee

Behind the Counter:
Machine　Dalla Corte Zero 3 Group
Grinder　Mahlkönig EK43, Anfim SP II 　　Roaster　Probat

Paying homage to one of the original Korean specialty coffee houses of Seoul, La Pluma & Bohemian was opened by industry veteran Choi Young-sook (최영숙) in 2021.

No history of Korea's specialty coffee industry would be complete without mentioning Bohemian Coffee's

and the brand's founder Park Yi-chu (박이추). Born in Kyushu (Japan) in 1950, Park continued his education in Tokyo before moving to Seoul permanently in 1988 to establish his first cafe "International Coffee House Bohemian" in the city's bustling university district Hyehwa-dong. At a time when the only coffee consumed in Korea was mixed instant coffee, his take on a Japanese style "kissaten" with its deep mahogany interior, Dutch drip coffee and over 150 types of roasted coffee beans offered a completely new perspective to the next generation of coffee drinkers.

For the first time, coffee drinkers were able to experience the complete coffee process in-house from roasting to brewing while being introduced to the different flavours of coffee from far-flung regions in Panama (Esmeralda Geisha), Jamaica (Blue Mountain), Yemen (Mocha Matta) and Cuba (Crystal Mountain).

Operating continuously from 1990 until October 2021, Park moved the business to the East coast city of Gangneung (강릉) in 2004, leaving the Anam-dong branch in the reliable hands of Choi. After running the Bohemian Coffee House for 17 years, she eventually closed the cafe, continuing the legacy with La Pluma just a short walk away near Korea University's vibrant back-gate.

In one of the most historic coffee events to take place in the country,

the opening of La Pluma was followed by an extraordinary pop-up of former Bohemian regulars, industry veterans and champion baristas. Among the hundreds that gathered for the occasion, Seo Pil-hoon (Coffee Libre), Kim Sa-hong (Coffee Temple), Kwon Yo-seop (Hell Cafe Roasters), Kim Byeong-gi (Fritz Coffee Company), Sa Seon-hui (Cyphon Coffee), Lee Seung-jin (180 Coffee Roasters) and Lee Dong-yeop (Escudo Coffee) all came together to pay homage to the brand and reminisce over thirty years of fond Bohemian memories.

Menu Old-school line-up of Dutch and Drip coffees with a choice of three different blends (Beethoven, Bach, and Hemingway), a daily special (Single Origin hand-drip) and two varieties of Panama Geisha.

Onion*

어니언

📍 5 Gyedong-gil (Gye-dong), Jongno-gu, Seoul

🚇 Anguk (Line 3, Exit 3)   📷 @cafe.onion

🗓 **Mon to Fri** 7:00am - 10:00pm / **Sat to Sun** 9:00am - 10:00pm

Established **2019**   Brewing Method **Espresso, Filter (V60)**
Recommended Menu **Americano / Italian Cappuccino**

Behind the Counter:
Machine **Synesso MVP Hydra 3 Group**
Grinder **Victoria Arduino Mythos One**   Roaster **Loring S7 Nighthawk**

This is the third of the Onion cafes in Seoul, located near some of the most visited tourist attractions in Northern Seoul; Bukchon Hanok Village (북촌한옥마을), Gyeongbokgung Palace (경복궁) and Samcheong-dong (삼청동). Be sure to tick them off after swinging by Onion for a caffeine pit stop and delicious morning pastry.
Designed by local artist studio

Fabrikr, the huge 661m² complex is built inside a renovated traditional Korean hanok building, complete with central courtyard, minimalist mahogany-decked lounge area and stylish floor-to-ceiling glass facades. Take a seat in the cafe's wide-open patio area and appreciate one of the most Korean settings imaginable for a coffee.

Baked goods include croissants, strawberry tarts, scones, pandoros, sandwiches and their special dark chocolate salty butter roll.

Concept Famous for re-generating existing spaces, Onion created created their Jongno branch inside a landmark hanok building. Come here for traditional vibes, freshly baked cakes and Onion's in-house roasted specialty coffee.

Other Branches:
- **Onion (Seongsu - 성수)**
  - 8 Achasan-ro 9-gil (Seongsu-dong 2-ga), Seongdong-gu, Seoul
  - Seongsu (Line 2, Exit 4)
- **Onion (Mia - 미아)**
  - 55 Solmae-ro 50-gil (Mia-dong), Gangbuk-gu, Seoul
  - Mia (Line 4, Exit 4)

Second Coffee

세컨드커피

📍 16 Yulgok-ro 10-gil (Waryong-dong), Jongno-gu, Seoul
🚇 Anguk (Line 3, Exit 4)　　📷 @secondcoffee
🏪 **Wed to Sun** 11:30am - 7:00pm / **Mon to Tue** Closed

Established **2021**　　Brewing Method **Espresso, Filter (V60)**
Recommended Menu **Con Panna**

Behind the Counter:
Machine **Faema E61 2 Group**　　Grinder **Mahlkönig EK43, E60, Ditting Peak**
Roaster **Probat 1969 UG-15**

Before opening Second Coffee in October 2021, owner, roaster and barista Kim Jung-hoi (김정회) trained for over a decade in a variety of cafe backgrounds across Seoul. After starting with Mussetti, one of the first Italian style espresso bars in Seoul, Kim worked at Coffee Themselves while participating in Korea National Barista competitions, making it to the

finals consecutively in the 2010 and 2011 competitions. In 2013 he started to focus on roasting and has been perfecting his skills ever since on a series of state-of-the-art Probat UG series roasters, one of which, a 1969 UG-15 model, sits proudly in his own roasting lab. As well as roasting for his own brand, Kim has an established network of partner cafes across the city who rely on Second Coffee's medium roasted beans, ideally suited for thick crema espressos and sweet con pannas.

Located in the heart of Seoul's traditional city centre, Second Coffee is just a stone's throw away from some of the capital's main historic attractions. Take a walk down nearby Seosulla-gil (서순라길), just over the wall from Chosun dynasty ancestral shrine Jongmyo (a UNESCO World Heritage site) and you can experience the city as it looked in the 15th and 16th centuries. Starting next to Exit 11 of Jongno 3-ga station (Line 1, 3 and 5), the narrow pedestrian only (weekends) laneway ends in front of Changdeokgung Palace (창덕궁), a sprawling royal palace and garden complex established in 1405.

Coffee Beans Second Coffee roasts up a combination of two house espresso blends Dusk and Night, along with one seasonal single origin for filter coffee.

Espresso Beans in Korea Before domestic brands like Second Coffee started roasting in-house, cafes relied almost entirely on imported Italian blends from the likes of Lavazza (Piedmont), Illy (Trieste) and Arcaffe (Livorno) for their espresso beans. While these Italian brands still exist in Korea, the market has changed

exponentially with many espresso bars now opting to roast their own and even invent their own fusion espresso recipes. Some examples of this recent espresso bar boom in Seoul include Leesar Coffee (리사르커피), Bamaself (바마셀) and MonsieurBuBu CoffeeStand (무슈부부 커피스탠드).

Interior Originally a local convenience store, Second Coffee's interior has been completely revamped with mahogany and beige colour tones, retro 1970s style lighting and a sleek stainless steel-topped espresso bar.

Nothin Coffee

노띵커피

📍 33-2 Toegye-ro 50-gil (Mukjeong-dong), Jung-gu, Seoul
🚇 Dongguk University (Line 3, Exit 6), Chungmu-ro (Line 3 and 4, Exit 1)
📷 @nothincoffee   📅 **Mon to Sat** 9:00am - 7:00pm / **Sun** Closed

Established **2016**   Brewing Method **Espresso, Filter** (Steadfast)
Recommended Menu **Single Origin Hand Drip / Espresso**

Behind the Counter:
Machine **Slayer Espresso 2 Group**
Grinder **Etzinger etzMAX, Mahlkönig E80, Ditting 807**   Roaster **Diedrich IR-5**

From the direct sourcing of high-quality green coffee beans, design of both their original Goyang-si (고양시) and Seoul cafes to the engineering of their own hand-drip brewer, Nothin Coffee founder Kim Hyun-joon (김현준) has been absorbed in this industry for over 20 years. It's taken him and his family around the world visiting coffee growing regions, taking Q-Grader sensory exams in the United States and travelling to some of the world's most famous cafes everywhere from Italy to Japan. Eventually they decided to settle down in Gyeonggi-do (경기도) with their first roastery and hand-drip coffee house opening in 2016, just a short walk from the picturesque Seooreung Royal Tombs (서오릉). Moving to central Seoul in 2021 the brand now occupies a renovated three storey building in Mukjeong-dong (묵정동). Only a few minutes from downtown Eulji-ro (을지로), this quiet neighbourhood retains much of its nostalgic charm with generation-old restaurants, 1970s-style villa residences and local grocery stores. The colourful building itself has an interesting array of features from its industrial arched metal window frames, Terrazzo marble chip-textured exterior and a vaulted terracotta brick interior ceiling. Seating is available on the first and third floors while the roasting lab and office take up the second floor. A further seating area on the rooftop opens between spring and autumn. Nothin Coffee's menu includes half a dozen rare single origins from far flung growing destinations like Yemen, Panama and Peru sold alongside decaf options and a range of traditional Chinese and Korean herbal teas. To pair, Nothin also bake their own desserts including cafe staples pound cake, lemon cake, cranberry scones and almond financiers.

Steadfast Dripper The family-run business have developed a number of their own products in the past, including an industry-first specialty instant coffee, convenient drip bags and even their own patented dripper called "Steadfast". Launched in August 2021, the stainless steel brewer, finished with a natural leather holder, uses a capillary extraction method similar to the Kalita Ceramic Dripper.

Sustainable Coffee As well as focusing on how to brew good coffee, Nothin Coffee makes a conscious effort to reduce the environmental impact of their business by using eco-friendly packaging and biodegradable bean bags ever since opening their first cafe in 2016.

Coffee @ Works

커피앳웍스

📍 Leema Building, 42 Jong-ro 1-gil (Susong-dong), Jongno-gu, Seoul
🚇 Gwanghwamun (Line 5, Exit 2)　　📷 @cw_coffeeworks
📅 **Mon to Fri** 7:00am - 9:00pm / **Sat to Sun** 8:00am - 7:00pm

Established **2014**　　Brewing Method **Espresso, Filter** (V60, Chemex, Siphon)**, Cold Brew**
Recommended Menu **Barista's Choice Brewed Coffee** (Filter)

Behind the Counter:
Machine **La Marzocco Strada EP 3 Group**
Grinder **Mahlkönig EK43, Ditting KE640, KED640**　　Roaster **Loring S35 Kestrel**

Recently renovated, Coffee @ Works' flagship branch is located on the ground floor of Leema (이마, 利馬) Building, an iconic skyscraper in Seoul's central business district just a short walk from Gwanghwamun (광화문) square. Covering over 300m², the huge showroom space is the perfect place to sit back and enjoy a cup of coffee in the heart of the city; choose from a sofa in the lounge area, grab a front row seat at the central brew bar or focus on some work in the private study booths. Established in 2014 by Korean food and beverage conglomerate, SPC Group, Coffee @ Works is the first domestic specialty coffee franchise to be launched in the country. Competing with overseas brands Starbucks and Blue Bottle, Coffee@Works focuses purely on specialty coffee and is limited to just six bespoke branches in Seoul. Despite the brand being less than 10 years old, parent company SPC Group's relationship with coffee goes back decades to the opening of their bakery-cafe franchise Paris Baguette in 1988. The instantly recognisable brand now has over 3,300 branches

across the country and more than 400 overseas stores including a recent venture into Paris itself, the "Paris Baguette Saint-Michel".

Coffee @ Works control almost every part of the process involved across their range of specialty coffee beans. Over 90% of the beans they use are sourced via direct trade channels, they roast in their own factory on a state-of-the-art Loring machine, and even have their own patented coffee fermentation process developed in collaboration with partner farms in Colombia.

Menu When it comes to ordering we recommend going for the "Barista's Choice" filter coffee, brewed using a variety of options from V60 to Chemex and even Siphon. As well as single origin brewed coffee, there is a full espresso-based menu using house blends Black and Blue, Diva and a de-caffeinated variety aptly named Nocturne. For take-home options choose from a selection of cold brew, drip bags and specialty capsule coffees.

In-house R&D In partnership with the SPC Group's Food Biotech Research Institute, the Coffee @ Works team have developed their own anaerobic fermentation technology using patented yeast and lactic acid bacteria. In a process that took over four years, working closely with El Paraiso Farm in Cauca, Colombia, they now apply this to their coffee crop to maximise the original flavour and produce a variety brew unlike any other in the market.

North East, East

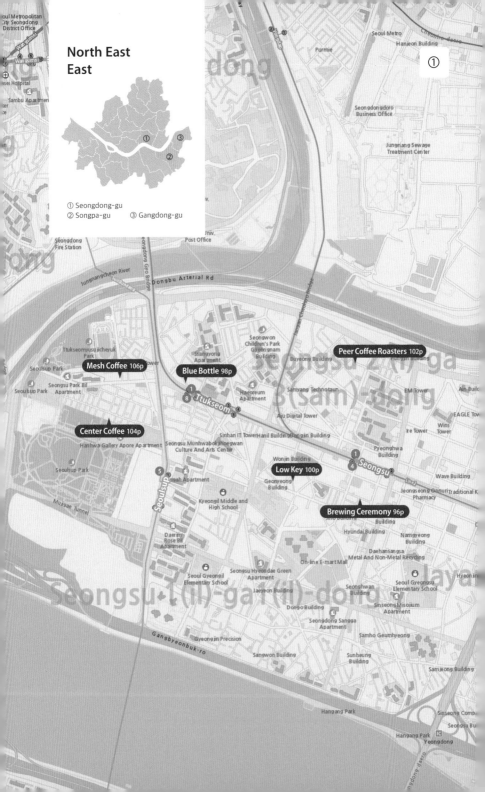

① Seongdong-gu
② Songpa-gu
③ Gangdong-gu

North East
East

① Seongdong-gu
② Songpa-gu
③ Gangdong-gu

Mesh Coffee 106p

Blue Bottle 98p

Peer Coffee Roasters 102p

Center Coffee 104p

Low Key 100p

Brewing Ceremony 96p

②+③

Coffee Montage 114p

Iwol Roasters 112p

Within Coffee 110p

Brewing Ceremony

브루잉 세레모니

📍 22-1 Yeonmujang 5ga-gil (Seongsu-dong 2-ga), Seongdong-gu, Seoul
🚇 Seongsu (Line 2, Exit 4)   📷 @brewingceremony
📅 **Mon to Sat** 11:00am - 9:00pm / **Sun** Closed

Established 2019   Brewing Method **Filter** (Kalita Ceramic), **Espresso**
Recommended Menu Single Origin Brewing

Behind the Counter:
Machine La Marzocco GS3 1 Group   Grinder Ditting KR804, Mazzer Robur
Roaster Probat Probatino

Another cafe curated by interior design powerhouse "The First Penguin", Brewing Ceremony is perhaps one of the most instantly recognisable cafes in Seongsu-dong. The compact street-level space has a space-age vibe with huge curved, tinted windows, art deco

stainless steel tables and a suspended monolith hanging at the front cave-like entrance.

Inspired by the meditative ambience of a Korean "darye" (tea ceremony or 다례 in Korean), owner and head roaster Choi Wan-seong (최완성), chose this former office location to create a deliberate contrast between the outside thoroughfare and Brewing Ceremony's peaceful interior. A rare respite from Seoul's busy streets, you can feel the tempo immediately slow down as you step off the alleyway, through the darkened glass doors and into this unique space.

Brewing Ceremony focuses naturally on filter coffee and produces small batches in-house on their custom roaster, grinding beans for each cup to perfection on their Ditting (Switzerland) KR804 electronic precision grinder. Past favourites and guest beans include Honduras La Montañita, Panama Geisha Boquete Yeni and Kenya Peaberry Oakland fully washed.

Other specialty cafe concepts designed by "The First Penguin" include txt Coffee (txt 커피), GreenMile Coffee Bukchon (그린마일 커피 북촌점), Parched Seoul (파치드 서울) and das ist PROBAT (다스 이스트 프로밧).

From Starbucks to Specialty Roaster
Owner and head roaster Choi Wanseong worked for Starbucks for over 17 years before a short stint studying roasting at Ilya Espresso (일야커피) in Seoul's Gongdeok neighbourhood. After a comprehensive cafe tour of Tokyo, Choi returned to Seoul and established his own business.

Blue Bottle

블루보틀

📍 7 Achasan-ro (Seongsu-dong 1-ga), Seongdong-gu, Seoul

🚇 Ttukseom (Line 2, Exit 1)   📷 @bluebottlecoffee_korea

🗓 **Mon to Sun** 8:00am - 8:00pm

Established **2019**   Brewing Method **Espresso, Filter (V60)**
Recommended Menu **New Orleans**

Behind the Counter:
Machine **La Marzocco Linea 2 Group**
Grinder **Mazzer Robur, Mahlkönig EK43**   Roaster **Loring S35 Kestrel**

Four years after opening their first Asian branch in Tokyo, Blue Bottle finally entered the Korean market in May 2019. In the years before, CEO Bryan Meehan made frequent appearances in Seoul, most notably at the Seoul Cafe Show, and regularly hinted on their intentions to open a local branch. It was common knowledge to many of the team that Koreans had long made up a significant number of the visitors in popular overseas destinations like San Francisco, Tokyo and Kyoto. On the first day Blue Bottle recorded an astonishing $50,000 of sales with up to 400 people in the queue before the doors had even opened on 3rd May 2019. For weeks, Instagram was awash with Blue Bottle hashtags, photos of goods and the long winding queues outside Korea's hottest cafe. The launch was so successful

it almost single-handedly changed the appearance and fortunes of the Seongsu-dong neighbourhood. Coined "the Blue Bottle effect" the area went through a rapid period of gentrification that saw one of the most concentrated growth in cafes that the city had ever seen.

The first branch remains their HQ while also serving as a roastery (one of only four such facilities for Blue Bottle worldwide) and training lab for the growing team of baristas.

In true Blue Bottle style the aesthetics of each venue blend seamlessly into the surroundings, from their traditional "Hanok" style building in Samcheong-dong (삼청동) to their sleek Yeouido (여의도) branch at the heart of the Hyundai Department Store. So confident of the demand for Blue Bottle coffee in Korea, the company has since invested in a total of 10 different locations across Seoul

and recently Jeju island.

Brand Localisation As well as carefully considering Korean aesthetics for each branch, Blue Bottle has successfully teamed up with local brands on a number of occasions, including the popular "Coffee Golden Ale" collaboration with the Jeju Beer Company.

Other Branches:
- **Blue Bottle (Gwanghwamun - 광화문)**
  - 11 Cheonggyecheon-ro (Seorin-dong), Jongno-gu, Seoul
  - Gwanghwamun (Line 5, Exit 5)
- **Blue Bottle (Samcheong - 삼청)**
  - 76 Bukchon-ro 5-gil (Sogyeok-dong), Jongno-gu, Seoul
  - Anguk (Line 3, Exit 2)
- **Blue Bottle (Yeouido - 여의도)**
  - 5th Floo, The Hyundai Seoul, 108 Yeoui-daero (Yeouido-dong), Yeongdeungpo-gu, Seoul
  - Yeouinaru (Line 5, Exit 1)

Low Key

로우키

📍 6 Yeonmujang 3-gil (Seongsu-dong 2-ga), Seongdong-gu, Seoul

🚇 Seongsu (Line 2, Exit 4) 📷 @lowkey_coffee

📅 **Mon to Fri** 10:00am - 8:00pm / **Sat** 12:00pm - 9:00pm / **Sun** 12:00pm - 6:00pm

Established **2010**   Brewing Method **Espresso, Filter (V60), Cold Brew**
Recommended Menu **Brewing**

Behind the Counter:
Machine **La Marzocco Strada Electronic Paddle EP ("Street Team")**
Grinder **Anfim SP II, Mahlkönig EK43**

Specialty cafe and small batch roaster founded in 2010, Low Key is a true trend setter on the Korean coffee scene. For the last decade they have kept the business to just two cafes, dedicating their time instead to wholesale distribution, coffee subscription services and coffee education through public cupping. More recently they've also increased their product offering to meet the exponential demand for "home-cafe" solutions including Capsule Coffee (for Nespresso), Coffee Drip Bags and Cold Brew.

Beans Low Key import green beans directly from source including Finca Deborah precious Geisha from the Chiriqui Volcan Region of Panama, as well as Burundi, Colombia and El Salvador Cup of Excellence (CoE) varieties.

Packaging Roasted beans are packaged in envelopes; a stamp for the country of origin, another stamp for the date of roasting and a post code number detailing the altitude of the farm.

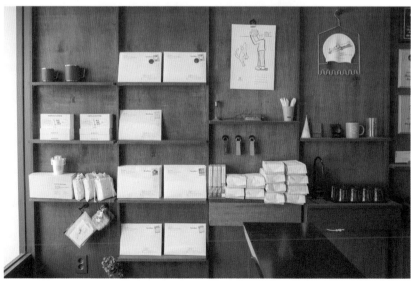

Peer Coffee Roasters*

피어커피로스터스

📍 24 Gwangnaru-ro 4ga-gil (Seongsu-dong 2-ga), Seongdong-gu, Seoul

🚇 Seongsu (Line 2, Exit 1)   📷 @peer_coffee_roasters

🗓 **Mon to Fri** 10:00am - 5:00pm / **Sat to Sun** 12:00pm - 8:00pm

Established 2015   Brewing Method **Espresso, Filter** (Kalita Wave)

Recommended Menu **Brewing Coffee**

Behind the Counter:

Machine **Victoria Arduino VA388 Black Eagle 2 Group**

Grinder **Victoria Arduino Mythos One, Mahlkönig EK43, Mazzer Robur, Ditting 807**

Roaster **Probat P05**

Established in 2015 in Hannam-dong, Peer Coffee Roasters moved to Seongsu-dong in 2019, opening a roasting workshop and showroom cafe in the quiet backstreets of Seoul's coffee capital. Slightly larger than their Dongdaemun branch the Seongsu-dong flagship cafe has up to eight single origins in the line-up offering the chance for fans of the brand to find a favourite before choosing beans from the take-home section.

Alongside signature espresso house blends "El Camino" (Colombia, Brazil and Guatemala) and "Diana" (3 different Ethiopian beans) this varied selection has attracted a lot of attention particularly during Covid-19 lockdowns as cafes moved to subscription-based sales. Peer Coffee's version, the "Home Barista Club", kept pushing the boundaries with some adventurous crops, all roasted in-house on Peer's precious Probat.

Other Branches:

• **Peer Coffee Roasters (Gwanghuimun - 광희문)**

📍 123 Cheonggu-ro (Sindang-dong), Jung-gu, Seoul

🚇 Cheonggu (Line 5 and 6, Exit 1), Dongdaemun History & Culture Park (Line 2, 4 and 5, Exit 3)

Center Coffee

센터커피

📍 28-11 Seoulsup 2-gil (Seongsu-dong 1-ga), Seongdong-gu, Seoul

🚇 Seoulsup (Suin-Bundang Line, Exit 5)   📷 @centercoffee

📅 **Tue to Sun** 10:00am - 9:00pm / **Mon** Closed

Established **2017**   Brewing Method **Espresso, Filter** (Clever Dripper)**, Cold Brew**
Recommended Menu **Espresso / Cold Brew Geisha**

Behind the Counter:
Machine **Kees van der Westen Spirit Triplette** (3 Group)   Grinder **Mahlkönig EK43, E80**
Roaster **Giesen W6, Stronghold S7, S9, Probat Probatino**

Opened by champion roaster Park Sang-ho (박상호) and actor Bae Yong-joon (배용준), Center Coffee is one of the main draws for coffee enthusiasts on Seongsu-dong's shopping and restaurant alleyway Seoulsup 2-gil (서울숲2길). Located just metres from the entrance to Seoul Forest Park, this trendy cafe is disguised in a renovated residential brick facade building. The industrial chic interior spreads out over a two-storey space complete with polished chrome counters, exposed steel beams and wide-framed factory style windows.

Originally based in the UK from high-school, head barista Park Sang-ho boasts an impressive resume. He was the overall champion at the 2013 UK Brewers Cup, won the 2015 Coffee in Good Spirits Championship and served as head roaster at James Hoffman's legendary Square Mile Coffee Roasters. From the high end Kees van der

Westen espresso machine to the multitude of roasters behind the scenes, Center Coffee aims to take Seoul Forest's coffee scene to another level. In the coffee department expect a rare combination of Geisha from Panama (Finca Deborah), Colombia (La Esperanza) and Ethiopia alongside seasonal blends and cold brews. Center Coffee can also provide for all of your home cafe needs too with capsules (decaf and normal), drip coffee bags and their own branded Mr. Clever coffee drippers.

Other Branches:

• **Center Coffee (Seoul Station - 서울역)**
  📍 4th floor, Seoul Station (KTX), 405 Hangang-daero (Dongja-dong), Yongsan-gu, Seoul
  🚇 Seoul Station (Line 4 and 1, Airport Line, Gyeongui-Jungang Line, Exit 1, Follow signs to KTX)

Mesh Coffee*

메쉬커피

📍 43 Seoulsup-gil (Seongsu-dong 1-ga), Seongdong-gu, Seoul

🚇 Ttukseom (Line 2, Exit 8)　　📷 @meshcoffee

📅 **Mon to Fri** 8:30am - 5:00pm / **Sat** 10:00am - 6:00pm / **Sun** 12:00pm - 5:00pm

Established **2015**　　Brewing Method **Espresso, Filter (V60)**
Recommended Menu **Flat White / Single Origin Brewing Coffee**

Behind the Counter:
Machine MOAI Bar System Two Group
Grinder Mahlkönig EK43, Anfim Super Caimano, Victoria Arduino Mythos One
Roaster Taehwan Proaster

Established by industry veterans Kim Hyeon-seob (김현섭) and Kim gi-hoon (김기훈) Mesh, after Fritz (프릳츠) and Coffee Libre (커피 리브레), is one of the most influential specialty cafe brands in Seoul. Combined with the arrival of Blue Bottle in 2019, the two cafes have gradually turned the eastern suburb of Seongsu-dong into one of the hottest cafe areas in the city.

Part of what makes Mesh so unique is the proximity of the cafe to their roasting lab next-door. In fact all that divides the two spaces is a single floor-to-ceiling glass wall; head roaster Hyeon-seob regularly serves coffee in between roasts, sharing knowledge with regulars, coffee bean customers and chatting with first-time visitors to the area.

Known by locals simply as "Mesh" the brand and its directors, the "Artistic Duo" of Hyeon-seob and Gi-hoon, have become synonymous with specialty coffee in the city. Their understanding of coffee, from the culture driving it, to their overseas network of roasters and green bean suppliers, is almost unparalleled in Korea. Head roaster Kim Hyeon-seob has a passion for Scandinavian light roasted coffee and over the years has developed a special

"It's just coffee after all"

relationship with some of the most important players in the Nordic region, including Tim Wendelboe (Norway), Drop Coffee (Sweden) and La Cabra Coffee (Denmark).

Having established the cafe brand the team now devote more time in educating future generations with regular publications, media interviews and overseeing National Barista Competitions. Their first book on the subject "Oh yeah! Specialty Coffee" written in 2018 was one of the first books to cover the subject in detail and continues to be a top seller in its category.

The Seongsu-dong branch is still thriving seven years on, and with the brand legacy in safe hands, Mesh recently opened up a second cafe over the river in Itaewon's "Freedom Village" (해방촌).

Micro-Roastery Pictured here is the ground floor lab and standing bar (only three seats!) where it all started in 2015.

Roasting Mesh founder Kim Hyeon-seob originally studied coffee roasting under Coffee Libre's Seo Pil-hoon (서필훈) from as early as 2010. Inspired by Libre's technical approach to specialty coffee he continued to study before opening the Seongsu-dong cafe in 2015.

Other Branches:
- **Mesh (HBC - 해방촌)**
  - 10 Sinheung-ro 7-gil (Yongsan-dong 2-ga), Yongsan-gu, Seoul
  - Noksapyeong (Line 6, Exit 2)

Within Coffee

위딘커피

📍 4-9 Garak-ro 17-gil (Songpa-dong), Songpa-gu, Seoul
🚇 Seokchon (Line 9 and 8, Exit 4), Songpa (Line 8, Exit 1)   📷 @withincoffee
📅 **Mon to Fri** 11:00am - 5:00pm / **Wed** Closed / **Sat to Sun** 11:00am - 6:00pm

Established 2020   Brewing Method Espresso, Filter (V60, Aeropress)
Recommended Menu Aeropress

Behind the Counter:
Machine Dalla Corte XT Barista 3 Group
Grinder Mahlkönig EK43, Victoria Arduino Mythos Two
Roaster Taehwan Proaster THCR-03

Established by former Mesh Coffee barista Song Han (송한), Within Coffee is a specialty cafe with a distinct homely vibe. Nestled in the quiet backstreets of Songpa, the cafe features a central breakfast bar style counter, mini roasting lab and outdoor stone patio, ideal for a leisurely lunchtime

cappuccino.

The coffee menu includes both filter (Aeropress) and espresso options, all using Within's in-house roasted coffee beans. Expect to see a daily rotating espresso menu alongside a selection of rare single origin filter options, many of them sourced through Oslo-based green bean supplier Nordic Approach. Having grown up in Finland, Song has been a long-time advocate for the lighter style of coffee now commonly associated with Scandinavia, regularly introducing new batches through public-cupping and collaborations with partner cafes.

Keep an eye on the special menu "Something Sweet" too, the highlight being "Mini Bom", a version of the Spanish "Café Bombón" (Spanish for confection) a half-half concoction of sweetened condensed milk and coffee.

Baked goods are also available including a Finnish style Cinnamon Roll (Korvapuusti).

Iwol Roasters

이월로스터스

📍 14 Baekjegobun-ro 45-gil (Songpa-dong), Songpa-gu, Seoul
🚇 Seokchon (Line 8 and 9, Exit 2), Songpanaru (Line 9, Exit 1)   📷 @february_roasters
🏪 **Mon to Sun** 9:30am - 10:00pm

Established 2014   Brewing Method Espresso, Filter (V60), Cold Brew
Recommended Menu December Latte

Behind the Counter:
Machine La Marzocco Linea 2 Group   Grinder Mahlkönig K30 Twin 2.0, Tanzania
Roaster Probat P25, Trinitas T2, IKAWA Pro V3 Sample Roaster

Starting from a modest 26m² roastery in 2014, owner and head-roaster Kim Jae-hyeob (김재협) has built an impressive coffee empire over the last eight years. Moving from Youngin (용인) in 2015, the business is now headquartered at their Maseok (마석) roasting factory with three cafes spread across Songpa-gu (송파구) and 3rd wave cafe hub Seongsu-dong (성수동).

Our favourite branch sits on the popular shopping and restaurant street Songridan-gil (송리단길), just a few minutes walk from iconic Seoul landmark Seokchonhosu Lake (석촌호수). Located on a busy intersection, Iwol (February or 이월 in Korean) Roasters is an oasis of calm amidst the hustle and bustle of this vibrant area of fusion restaurants, trendy bars and traditional Korean eateries.

Entering the cafe, the pentagon-shaped space is divided down the middle by two large shared workbenches and a central coffee counter. Surrounded by comforting tones of silver, white and mahogany, this is the ideal setting to enjoy a quiet brew before

heading out to the bright lights and city sights of Seoul's largest district. Minimalist suspended lighting runs the length of each table providing a library-like backdrop, while the cafe's floor-to-ceiling windows are carefully glazed to minimise distraction and keep the interior as relaxed as possible. Originally having roasted on a single Diedrich IR-5, Kim and his team now roast batches of up to 25kg at a time on their high-end line up of Probat, Trinitas and IKAWA machines. Focusing on small-to-medium batch single origins, Iwol Roasters typically offer several

different filter options at any one time, alongside house blend "Manwol" (Full Moon), an eclectic combination of Colombian, Kenyan, Guatemalan and El Salvador beans. Keep an eye out too for menu staples, Einspanner, December Latte (almond syrup base), April Cream Latte and peppermint-lime infused June Ade.

Other Branches:
- **Iwol Roasters (Seongsu - 성수)**
  - 📍 2nd Floor, 46 Seoulsup 2-gil (Seongsu-dong 1-ga), Seongdong-gu, Seoul
  - 🚇 Ttukseom (Line 2, Exit 8), Seoulsup (Suin-Bundang Line, Exit 5)

Coffee Montage*

커피몽타주

📍 23-12 Olympic-ro 48-gil (Seongnae-dong), Gangdong-gu, Seoul
🚇 Gangdong-gu Office (Line 8, Exit 3) 📷 @coffeemontage
🕐 **Mon to Fri** 8:00am - 6:00pm / **Sat to Sun** 10:00am - 8:00pm

Established 2013   Brewing Method Espresso, Filter (V60), Cold Brew
Recommended Menu Espresso Platter (Espresso + Macchiato) / Cold Brew Sampler

Behind the Counter:
Machine Dalla Corte Zero 3 Group   Grinder Mahlkönig EK43, Mazzer Robur, Anfim SP II
Roaster Loring S35 Kestrel, Probat Probatone 5

A key player in the Korean specialty coffee roasting scene, Coffee Montage has a reputation for great espressos, in-house roasted blends and innovative cold brew menus.

Now approaching ten years of business, the brand was originally set up as a small batch roaster in 2013 by Seoul-based husband and wife team Shin Jae-woong (신재웅) and Kwon Mi-jung (권미정). The core of the business remains focused on roasting and to this day the company operates out of a single flagship branch in the city's western suburb of Gangdong-gu. Roasted off-site in nearby Hanam City their coffee is served to hundreds of wholesale clients as well as a growing number of coffee bean subscribers through their aptly named "Weekly Montage" delivery service.

Recently re-modelled in 2020, the interior features a sleek marble and stainless steel bar, with benches and display shelving along the walls for brewing manuals, magazines and branded goods. The small venue lies at the heart of the community and often overflows during weekday lunchtimes, with queues of local office workers, students and regulars lining the adjacent alleyways.

Coffee Beans Coffee Montage's range of beans include "A Bitter-Sweet Life" Editions 1 and 2, "Sense and Sensibility", alongside an ever-changing range of seasonal single origins and Geisha auction lots. For an all rounded taste experience of Coffee Montage's menu, try their Espresso Platter or Cold Brew Sampler (Ice Cappuccino, Affogato and Cold Brew all for 9,000 KRW!).

North West, West

North West
West

① Mapo-gu
② Seodaemun-gu
③ Eunpyeong-gu

①+②

Kalas Coffee 146p

Reissue Coffee Roasters 148p

MonsieurBuBu 132p

Deep Blue Lake 134p

Cuppacity 126p

Cafe Comma 122p

Blot Coffee 120p

Pastel Coffee Works 128p

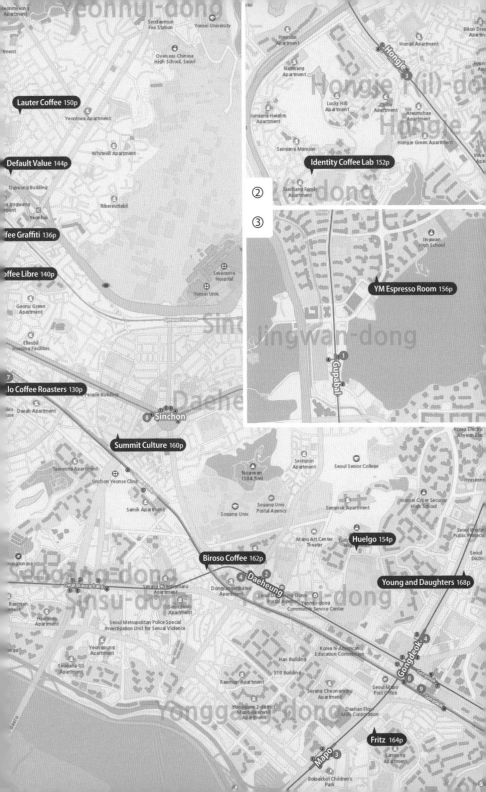

Lauter Coffee 150p

Default Value 144p

ffee Graffiti 136p

offee Libre 140p

lo Coffee Roasters 130p

Sinchon

Summit Culture 160p

Biroso Coffee 162p

Identity Coffee Lab 152p

②

③

YM Espresso Room 156p

Huelgo 154p

Young and Daughters 168p

Fritz 164p

Blot Coffee

블로트커피

6-5 Wausan-ro 19-gil (Seogyo-dong), Mapo-gu, Seoul
Hongik University (Line 2, Exit 9), Sangsu (Line 6, Exit 1)   @blot_coffee
Tue to Sun 12:00pm - 8:00pm / **Mon** Closed

Established 2018   Brewing Method Espresso, Filter (V60)
Recommended Menu Single Origin Filter Coffee

Behind the Counter:
Machine La Marzocco Linea 2 Group   Grinder Mazzer Robur, Mahlkönig EK43
Roaster Giesen W6

Established by the same team behind Zombie Coffee Roasters (2013 - 2018), Blot Coffee is one of the longest running coffee shops in the Hongdae neighbourhood. Along with brands

like Hell Cafe (헬카페), Felt (펠트) and Fritz (프릳츠), head roaster Lee Dae-ro (이대로) has roasted in-house since day one, travelling the world to establish relationships with growers and carefully selecting micro-lots to bring the freshest coffee possible back to Seoul.

Using their trusted Giesen W6, they roast the range of coffee beans including House Blend (Stalker) and Dark Blend (Ghost) alongside an ever growing range of specialty small batch lots, sourced directly from farmers in Panama, Honduras and Guatemala.

Cafe Comma

카페꼼마

📍 49 Poeun-ro (Hapjeong-dong), Mapo-gu, Seoul

🚇 Hapjeong (Line 6 and 2, Exit 8), Mangwon(Line 6, Exit 2)

📷 @cafecomma__official 🕐 **Mon to Sun** 10:00am - 10:00pm

Established **2011**

Brewing Method **Espresso, Filter (Poursteady Hario Brew Bar), Cold Brew**

Recommended Menu **House Blend Americano (Prime, Night Train)**

Behind the Counter:

Machine **La Marzocco Linea 3 Group**   Grinder **Mazzer Robur, Mahlkönig EK43**

Roaster **Probat P05, P25**

Study cafe with a growing focus on specialty coffee, Cafe Comma was first opened by Paju-based publishing giant Munhakdongne (문학동네) in 2011. Celebrating a decade of business last year, the cafe had a complete facelift, courtesy of renowned facade architect and city planner Yoo Hyun-joon (유현준).

Spread across six themed floors, this book lovers paradise includes two dedicated library levels, in-house coffee and baking labs as well as a rooftop terrace boasting panorama views of the surrounding Mangwon-dong (망원동) and Hapjeong-dong (합정동) neighbourhoods.

With the focus steadily increasing on specialty coffee, Cafe Comma decided to up their game in 2020 with the establishment of their own in-house roasting team, coffee lab and team of talented champion baristas. Instead of sourcing OEM branded coffee for the business, the team now run an

impressive operation using a pair of Probat P05 and P25 roasters to crank out seasonal single origins and house espresso blends "Prime" (Ethiopia, Colombia and Brazil) and "Night Train" (Honduras, Brazil). Paired with Cafe Comma's selection of freshly baked pastries, from croissants, almond financiers and chocolate cookies, this is the perfect place to spend a few hours browsing the bookshelves and indulging in a sweet bite to eat. Celebrating their 11th year in 2022, Cafe Comma continues to expand most recently with a branch in Seoul's Yeouido (여의도) suburb. Other branches worth exploring include Songdo (송도) and Sinsa-dong (신사동), the latter featuring a collaboration patisserie selection from world famous French pastry chef Yann

Couvreur (얀 쿠브레), the first of its kind for any cafe in Korea.

Book Club Cafe Comma runs a members book club in partnership with publisher Munhakdongne, offering monthly book recommendations, 50% discounts on coffee and exclusive invitations to events.

Other Branches:
- **Cafe Comma (Sinsa-dong - 신사동)**
  📍 Ground Floor, Sinsa Square, 652 Gangnam-daero (Sinsa-dong), Gangnam-gu, Seoul
  🚇 Sinsa (Line 3, Exit 6)
- **Cafe Comma (Songdo - 송도)**
  📍 263 Central-ro (Songdo-dong), Yeonsu-gu, Incheon
  🚇 International Business District (Incheon Line 1, Exit 5)

Cuppacity

커퍼시티

📍 48 Donggyo-ro (Hapjeong-dong), Mapo-gu, Seoul
🚇 Mangwon (Line 6, Exit 2)   📷 @this.is.cuppacity
🏬 **Mon to Wed** 11:00am - 6:00pm / **Thur to Sun** 8:00am - 6:00pm

Established **2021**   Brewing Method **Espresso, Filter (V60)**
Recommended Menu **Filter Coffee**

Behind the Counter:
Machine **La Marzocco Linea 2 Group**
Grinder **Mahlkönig X54, EK43**

Owner and barista Byeon Sang-hyun (변상현) first started in the coffee industry during a gap-year in Melbourne before returning to Seoul and working with nearby roastery Deep Blue Lake. Soon after, in early 2020, the perfect spot on Mangwon-dong's main high street turned up and Sang-hyun has been looking after the local's ever since.

At the heart of this friendly spot sits a long beech wooden countertop,

running almost the whole length of the cafe's street facing facade. Doubling as both the barista's work space and a breakfast bar style counter, this is the perfect place to sit down and plan out a day's sightseeing while enjoying a smooth flat white, Melbourne style short black or a fresh filter brew.

Instead of roasting on-site, Cuppacity uses a combination of local brands and award winning specialty beans from some of the world's favourite coffee cities. Recent line-ups include Market Lane Coffee (Melbourne), Red Brick Coffee (Canberra), Coffee Collective (Copenhagen) and Solberg & Hansen (Oslo).

Pastel Coffee Works

파스텔커피웍스

Ⓠ 38 Dongmak-ro 2-gil (Hapjeong-dong), Mapo-gu, Seoul

🚇 Hapjeong (Line 2 and 6, Exit 7)　　📷 @pastelcoffeeworks

🕙 **Mon to Fri** 8:30am - 5:00pm / **Sat to Sun** 10:00am – 6:00pm

Established **2011**　　Brewing Method **Espresso, Filter (V60), Cold Brew**
Recommended Menu **House Blend Drip Coffee**

Behind the Counter:
Machine **La Marzocco Linea 2 Group**　　Grinder **Mahlkönig EK43S, EK43**
Roaster **Taehwan Proaster**

Artisan small batch roaster Pastel Coffee Works recently celebrated its 10 year anniversary, a massive occasion for any cafe, let alone one in the heart of Korea's most competitive cafe suburb Hapjeong. Led by legendary barista, latte art champion and published coffee writer Jackie

"Experience the perfect cup"

Chang (장현우), the brand has grown exponentially and is now a significant player in the local B2B roasted coffee bean distribution industry.

Drawing from valuable connections made over years of attending World Barista Championships (since 2008) and travelling the world through cafes, Jackie is unrivalled in his knowledge of cafe destinations like Melbourne, Berlin and Copenhagen.

Keen to share his experience with the next generation, he writes regularly for local magazine *Monthly Coffee* as well as judging and taking part in Korea National Barista Championships (KNBC). Jackie is one of the longest standing competitors in this prestigious event, placing highly on a number of occasions including first place at the 2009 competition.

What started off as a mere 33m² cafe now hosts a barista training school, roasting factory and wholesale retail office along with a dozen staff. Roasting on-site using Korean-manufactured Taehwan Proaster machines, the team at Pastel are renowned nationwide for their creative blends, with classics like Lolly (the "Championship Blend") and cold brew Berliner setting the bar high for others in the industry to follow. If you're unsure what to order, check the "Today's Drip Coffee" board for their seasonal selection of single origins and famous house blends.

Millo Coffee Roasters[*]

밀로커피 로스터스

📍 36 Yanghwa-ro 18 an-gil (Donggyo-dong), Mapo-gu, Seoul
🚇 Hongik University (Line 2, Airport Line, Gyeongui-Jungang Line, Exit 7)
📷 @millocoffeeroasters   🏬 **Mon to Sun** 12:00pm - 10:00pm

Established **2008**   Brewing Method **Espresso, Filter (V60)**
Recommended Menu **Montblanc (Vienna Coffee)**

Behind the Counter:
Machine Synesso MVP 2 Group
Grinder Mazzer Robur, Victoria Arduino Mythos One, Etzinger etzMAX 2   Roaster Probat

A true first generation cafe, Millo has been entertaining students in one of the most cafe-populated suburbs of the city, Hongdae, for almost 20 years. Established in the early 2000s by industry veteran Hwang Dong-gu (황동구) this local spot serves the winning recipe, their legendary

"comfort coffee" the Montblanc (몽블랑). Similar to a Viennese Coffee, Millo created their version placing sweet homemade whipped cream on top of freshly pulled blended espresso. Many in the cafe industry now credit Millo for starting the Einspänner trend that continues to grow in popularity across both chain and independent cafes in the city.

As well as inventing new variation milk coffees, Millo Coffee Roasters have been at the forefront of testing high-end coffee machinery since the early days of specialty cafes in Korea. They were among the first to install

the La Marzocco GB5 as their espresso machine of choice in 2014, kick starting a revolution that would cement La Marzocco as the number one brand in the country. More recently, as well as changing to a Synesso, they have experimented with almost every brand of grinder, from Mazzer, Etzinger and Victoria Arduino to find the perfect match for their coffee.

Millo roast a signature espresso blend "Walking Stick" on their Probat roaster which features alongside a long line-up of small batch single origins available for hand drip filter coffee and americano.

Interior Inspired by the old coffee houses of Tokyo, or kissatens, Millo Coffee Roasters' interior features a long mahogany counter lined with retro drippers, antique brewing equipment and over 100 different cups displayed neatly behind the bar.

Hwang Dong-gu A legend of Korea's first generation of coffee brewers, owner and head roaster Hwang Dong-gu (황동구) has been perfecting his craft since the early 1990s. Come to Millo to witness one his perfect pour overs, just one of many ways to enjoy your time at this Hongdae institution.

MonsieurBuBu CoffeeStand

무슈부부 커피스탠드

📍 13 Mangwon-ro (Mangwon-dong), Mapo-gu, Seou

🚇 Mangwon (Line 6, Exit 2)   📷 @monsieurbubu.coffeestand

🏪 **Mon to Sun** 12:00pm - 8:00pm

Established **2018**

Brewing Method **Espresso, Dutch Coffee**

Recommended Menu **Lemon Romano Espresso / Con Panna**

Behind the Counter:

Machine **Nuova Simonelli Aurelia 2 Group**   Grinder **Anfim SP II**

Hole-in-the-wall espresso hot spot, MonsieurBuBu CoffeeStand is straight out of Europe with its vintage vibe, casual atmosphere and strong Italian espresso culture. This small alleyway space, carefully curated by designer and owner Kwon Oh-Hyeon (권오현), is the perfect place to unwind, just a block away from Hapjeong's bustling World Cup Street (월드컵로). Pop in for a

quiet espresso during the day or hang with the locals till late at the cafe's charming al fresco standing bar.
The style of coffee here is heavily influenced by Italy with classics Espresso Romano, Ristretto, Espresso Con Panna, Cortado and Dopios all featuring on the menu. We recommend going for the Espresso Romano (or a caffe al limone as it's known in Naples), an intriguing concoction made using a single espresso shot, one lemon slice and a teaspoon of homemade vanilla sugar. First, simply sip the espresso, then place the lemon slice in the cup, stir in the sugar and

enjoy. Wash down with a glass of sparkling water and charge your glass with a Con Panna or Americano to finish.
As well as coffee, MonsieurBuBu serves a selection of herbal teas, soft drinks and classic cocktails including High Balls (choice of Suntory or Jameson base), Bourbon-based Godfathers and caffeine-charged Espresso Martinis.

Coffee Couple Established in 2018 by husband and wife Kwon Oh-Hyeon and Park Seon-Young (박선영), the brand name "BuBu" (부부) is a nod to the Korean word for married couple.

Deep Blue Lake

딥블루레이크

📍 11 Poeun-ro 6-gil (Mangwon 1-dong), Mapo-gu, Seoul   🚇 Mangwon (Line 6, Exit 2)
📷 @deepbluelakecoffee   🕐 **Mon to Sun** 11:00am – 9:30pm

Established **2016**   Brewing Method **Espresso**
Recommended Menu **For Coffee Lover (One Espresso, One Latte and Sparkling Water)**

Behind the Counter:
Machine **Kees Van der Westen Spirit Triplette (3 Group)**
Grinder **Mazzer Kold S, Mahlkönig EK43**
Roaster **Loring S15 Falcon**

Over the course of the last five years, dozens of cafes have entered the re-developed Mangwon marketplace area, but few have gained a following quite like Deep Blue Lake. Established in 2016 by barista Lee Chul-won (이철원), the cafe's distinctive sky blue brick facade and symmetrical box windows have become a symbol of the neighbourhood and a popular starting point for any Mangwon-dong cafe tour. The repurposed residential building includes a lively coffee bar, office and goods shop on the ground floor, with plenty of seating available on the second and third floors. Each room is brightly decorated in Santorini blue, homely furnishings and views facing the Mangwon marketplace.

Deep Blue Lake has a constantly changing selection of single origin coffee beans alongside two house blends Blue and Deep, each featuring an exquisite combination of Ethiopian, Central American and Indian beans. Brand director and barista Lee Chul-won began studying coffee under Coffee Libre's Seo Pil-hoon (서필훈) in the early 2010s. Their relationship continues to this day as most of the green beans are sourced through Coffee Libre's distribution network.

Built on a reputation for high-end specialty coffee, the team pride themselves on their light "tea-like" Nordic roasting style, achieved using USA-manufactured roaster brand Loring. Beginning in 2014 with Coffee Graffiti, their Smart Roaster series (S15 Falcon and S35 Kestrel) are now widely used in Korea by popular roasting brands including Coffee Montage, Four B and Coffee Themselves.

You can try one of each of their coffees with their "For Coffee Lover" menu, which includes one espresso, one latte and a complimentary sparkling water. Please note the menu at Deep Blue Lake is composed entirely of espresso-based coffee and no brewing options are available. Beans are available to buy in-store though so feel free to take away and experiment with your next home brew.

Coffee Graffiti

커피 그래피티

📍 278 Donggyo-ro (Yeonnam-dong), Mapo-gu, Seoul
🚇 Hongik University (Line 2, Airport Line, Gyeongui-Jungang Line, Exit 3)
📷 @coffeegraffiti   🗓 **Mon to Sun** 12:00pm - 9:00pm

Established **2013**   Brewing Method **Espresso, Filter (V60)**
Recommended Menu **Single Origin Brewing Coffee (Geisha)**

Behind the Counter:
Machine **Victoria Arduino VA388 Black Eagle 3 Group**
Grinder **Victoria Arduino Mythos One, Ditting KR804, Mahlkönig EK43, E80**
Roaster **Loring S15 Falcon**

Now located in Seoul's caffeine capital Yeonnam-dong (연남동), Coffee Graffiti has been a gathering spot for coffee connoisseurs since first opening in nearby Seogyo-dong (서교동) almost 10 years ago.

Marked by a brightly lit "Yeonnam Graffiti" neon sign, the cafe sits inconspicuously on the ground floor of a commercial building near the northern perimeter of the suburb. Inside, the sophisticated white interior features a central counter, futuristic portal windows and plenty of group seating options spread throughout the open-plan area.

When it comes to ordering, you have the option of selecting from a long list of espresso varieties which can be paired with standard single origins or Panama Geisha varieties. Naturally, Geisha varieties come with a hefty price tag so make sure you take note before splashing out! CEO Lee Jong-

hoon (이종훈) imports premium coffee beans that few others in the country can access from the likes of Finca Deborah (Panama), QIMA (Yemen) and Kotowa Coffee (Panama). Famous for his stringent quality control, Lee only approves coffee beans after a minimum of three consecutive cuppings, often travelling to the growing regions in-person to build relationships with producers, discover new beans and find out the latest auction trends.

Lee first started as a barista in 2002 and has since become one of the most recognisable figures in South Korea's coffee industry. He has represented the country a total of three times at the World Barista Championship in 2004, 2009 and 2015, finishing 5th overall in his last bid in Seattle, Washington. After almost 20 years of competing on the global competition circuit, Lee recently turned his attention to writing and published his first essay, "Coffee Variety" during the Covid-19 lockdown of 2021. The first in a series of planned publications under the aptly named "Coffee Algorithm" brand, this incredibly detailed book contains up-to-date catalogues of existing wild, Ethiopian, Yemen and hybrid coffee subspecies.

Interior The ceiling is lined with painted white, exposed steel beams which cleverley display thousands of aluminium cans from the archives of Coffee Graffiti's vast collection of coffee beans.

Menu The brewing coffee menu here reads longer than your average cocktail bar. Alongside a standard americano and espresso menu, they consistently showcase some of the finest coffees available in the Korean market. Expect Panama Finca Deborah, Cafe Kotowa and Lamastus Family small batches from the Chiriqui region produced in experimental methods from Yeast Washed, Slow Dried and Washed Jugo processes.

Roaster Coffee Graffiti was the first company in Korea to start using the premium Loring (USA) brand of roasters after importing their first machine in 2014, a Loring S15 Falcon fluid-bed (2.5 - 15kg capacity).

Logo CEO and head roaster Lee Jong-hoon's theoretical approach to coffee is legendary in Korea. Even the logo, at first glance simply the letters C and G, is in-fact the cross section of a coffee bean.

Coffee Libre*

커피 리브레

📍 200 Seongmisan-ro (Yeonnam-dong), Mapo-gu, Seoul
🚇 Hongik University (Line 2, Airport Line, Gyeongui-Jungang Line, Exit 3)   📷 @coffeelibre_yn
📅 **Mon to Sun** 11:00am - 8:00pm

Established **2009**   Brewing Method **Espresso, Filter (V60)**
Recommended Menu **Americano**

Behind the Counter:
Machine **La Marzocco Linea 2 Group**   Grinder **Anfim SP II, Mazzer Robur S**
Roaster **Probat G90, UG45**

Head down the alleyways of Yeonnam-dong's Dongjin Market and you'll find one of the most influential cafes in Korea's specialty coffee scene. Founded by industry legend Seo Pil-hoon (서필훈), Coffee Libre is largely considered to be the first "third wave" cafe in the country. Since 2009, Seo is credited with bringing in a whole host of new coffees and specialty beans to the Korean market, all imported via the direct trading links he has built with plantations around the world. On average the brand imports 1,600 Tonnes of coffee annually from over 150 coffee farms in 12 countries. As well as selling online and in-store, they also operate a nationwide distribution business providing coffee to approximately 400 cafes in Korea. Despite a steady flow of new brands entering the market, Libre to this day maintains its position as the foremost green bean importer in the market. As we count the ever growing number of baristas, green bean traders and roasters in the country, it is astonishing to realise just how many trained under Libre or indeed worked alongside Seo at some point in their career. Still operating out of the same shop front, the small but welcoming original Libre is ideal for any coffeeholic looking for an authentic Korean coffee experience. Given the back of the cafe has one only table and a handful of counter seats, please be prepared to opt for a take-away coffee at this venue. There's plenty to see round the corner though, with

dozens of independent restaurants, craft markets and shops lining this lively alleyway.

As well as the Korean business, Coffee Libre runs a coffee farm in Nicaragua, and overseas cafes in Guatemala (opened 2017) and Shanghai too.

Early on in Seo's career he was taken under the wing of Kyoto-based Yuko Itoi of Times Club, who helped the former Bohemian Coffee apprentice make key connections with coffee plantations, roasters and overseas market players. He studied constantly, translating foreign roasting manuals, visiting North American trade shows and taking part in global cupping competitions.

Coffee Libre(Yeongdeungpo)

House Blends Three of the most popular blends on sale at Coffee Libre include Bad Blood (Bright & Syrupy), No Surprises (Sweet & Balance) and Dark Libre (Deep & Full).

Logo The inspiration for the name and company mascot came from the popular Jack Black film, Nacho Libre (2006).

Founder Seo Pil-hoon Captivated by his experience working at Bohemian Coffee House (Seoul, 1990 - 2021), apprentice Seo Pil-hoon (서필훈) would go on to create Korea's most successful green bean and roasted coffee distribution company "Coffee Libre". After working for four years in the Anam-dong branch he left Bohemian to travel the world, register as an SCAA Q-Grader in 2007 (the first ever Korean to do so) and eventually win two consecutive World Roasters Cups in 2012 and 2013.

Historic Setting Directly opposite the towering Times Square mall, Coffee Libre's Yeongdeungpo branch is located inside one of the oldest commercial buildings in the country. Once occupied by textile firm Kyungbang Limited, this iconic brick building has been on the same site since the mid-1930s.

Other Branches:

- **Coffee Libre (Yeongdeungpo - 영등포):**
  - Times Square Mall, 15 Yeongjung-ro (Yeongdeungpo-dong 4-ga), Yeongdeungpo-gu, Seoul
  - Yeongdeungpo (Line 1, Exit 6)
- **Coffee Libre (Myeongdong Cathedral - 명동성당):**
  - Myeongdong Cathedral, 74 Myeongdong-gil (Myeongdong 2-ga), Jung-gu, Seoul
  - Euljiro 3-ga (Line 2, Exit 12)
- **Coffee Libre (Gangnam Shinsegae - 강남 신세계):**
  - Basement Floor 1, Famille Street, 176 Shinbanpo-ro (Banpo-dong), Seocho-gu, Seoul
  - Express Bus Terminal (Line 3, 7 and 9, Exit 4)

Default Value

디폴트밸류

📍 333 Seongsan-ro (Yeonhui-dong), Seodaemun-gu, Seoul
🚇 Hongik University (Line 2, Airport Line, Gyeongui-Jungang Line, Exit 3)
📷 @defaultvalue_yeonhui   📅 **Mon to Sun** 11:00am - 8:00pm

Established **2020**   Brewing Method **Espresso, Filter (Siphon)**
Recommended Menu **Brewing Coffee (Siphon)**

Behind the Counter:
Machine **Faema E61 2 Group**
Grinder **Mazzer Robur, Anfim SP II, Mahlkönig EK43**   Roaster **Diedrich IR-5**

Default Value, with its monochrome interior and location facing the main Seongsan street is at the heart of one of the largest specialty coffee areas in Seoul, Yeonhui-dong.

Head barista and owner Shin Chang-ho (신창호) is a four time finalist of the Korea National Barista Championship (KNBC) and winner of the Korea Siphonist Championship (KSC) in

2015. Most recently, Shin took first place in the KNBC main event, held in June 2022. Aside from running the cafe, he also dedicates his time to educating the next generation of coffee professionals via his YouTube channel and barista training courses in roasting, cupping and Siphon brewing methods.

Default Value roasts in-house on their fire engine red iR-5 Series Diedrich (German-American) roaster, proudly on show at the rear end of the cafe. Kept busy with a constantly changing line up of single origins, expect unique beans from a wide spectrum of coffee

plantations.

Recent crops include Guatemala Santa Felisa Orange Honey, Colombia El Paraiso Lychee (Double Anaerobic Process) and Panama Don Julian Geisha Natural.

Ediya Coffee Both baristas Shin Chang-ho and Yoon Do-yeon (윤도연) started their career in coffee at the popular Korean chain Ediya Coffee (이디야커피). The brand, established in 2001, is by far the strongest local competitor to Starbucks and currently boasts twice as many branches as the American giant.

Mastering one of the more challenging (and dramatic!) methods of brewing coffee, barista Shin is one of the most experienced siphon brewers in Korea.

Kalas Coffee

칼라스커피

📍 16 World Cup buk-ro 16-gil (Seongsan-dong), Mapo-gu, Seoul

🚇 Hongik University (Line 2, Airport Line, Gyeongui-Jungang Line, Exit 3)   📷 @kalascoffee

📅 **Mon to Fri** 10:00am - 6:00pm / **Sat** 12:00pm - 6:00pm / **Sun** Closed

Established 2010   Brewing Method **Espresso, Filter (V60), Cold Brew**

Recommended Menu **Brewing Coffee, Seongsan Latte**

Behind the Counter:

Machine **Kees van der Westen 1 Group "Speedster"**

Grinder **Victoria Arduino Mythos One, Mazzer Robur, Mahlkönig EK43S, and PK100 Compak**

Roaster **Probat Probatino, P05**

Mapo-gu's roasting and specialty coffee institution Kalas Coffee, was inspired over a decade ago by legendary coffee entrepreneur Choi Min-geun (최민근). Nestled in the quiet backstreets of Seongsan-dong, this is an ideal spot for those looking to escape the crowds of Yeonnam-dong while enjoying an original selection of signature lattes and carefully brewed hand-drip coffees.

Newly renovated, their flagship showroom features an all-white laboratory style interior, V60 filter-inspired standing tables and on-site roasting factory.

Under the slogan "Enjoy your colourful life", Choi and the team focus on expressing coffee from every corner of the flavour spectrum. Roasting on premium German-made Probat machines since day one of the business in 2010, the team maintains a high standard of taste profiles across a range of single origins and seasonal blends. So intent on finding the perfect

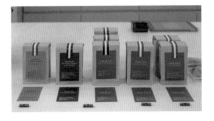

coffee, the passionate team conducts extensive green bean research on over 500 different coffees every year! As well as roasting wholesale, selling online and running the Seoul cafe, Kalas Coffee also operates an international branch in Leipzig, Germany. Long before establishing his name overseas, Choi was already a household name on the world coffee circuit, placing a respectable 3rd in the World Coffee Roasting Championship 2015, held in Gothenburg, Sweden. In the same year he beat fierce competition at home to be crowned Korean Roasting Champion and has since been on the judging panel for a whole host of domestic barista competitions.

Menu Focusing on delicate sweet notes and natural aromas, the baristas work with a huge selection of interesting specialty beans before selecting their final line up.

Reissue Coffee Roasters

리이슈 커피 로스터스

📍 35 Yeonnam-ro (Yeonnam-dong), Mapo-gu, Seoul
🚇 Hongik University (Line 2, Airport Line, Gyeongui-Jungang Line, Exit 3)   📷 @reissue.coffee
🕐 **Tue to Thur, Sun** 11:00am - 6:00pm / **Fri to Sat** 11:00am - 7:00pm / **Mon** Closed

Established **2015**   Brewing Method **Espresso**
Recommended Menu **Espresso (Medium Blend with Vanilla Sugar)**

Behind the Counter:
Machine La Marzocco Linea 2 Group
Grinder Anfim SP II, Mazzer Robur, Anfim Super Caimano
Roaster Diedrich IR-5, Taehwan Proaster Sampler

Seoul is brimming with espresso bars; from the classic Naples-style Leesar Coffee (리사르커피), to "standing room only" Bamaself (바마셀) and al fresco style MonsieurBuBu CoffeeStand (무슈부부 커피스탠드), the variety on offer is extraordinary. As more and more local coffee consumers start to enjoy the taste, aroma and variety of coffees available, the number of espresso bars has started to explode over the last two to three years.

Leading the trend for seven years now, Reissue Coffee Roasters has long since held its place as the number one espresso spot in Seoul's cafe capital Yeonnam-dong. Brightly coloured throughout with bold stripes of orange and white, Reissue's contemporary interior has a skate shop vibe to it with tiered "Bleacher-style" seating, sticker-covered walls and flowing hip-hop tunes.

Reissue's house blends, simply named

Medium and Dark, are the combination of over a decade of experience in the espresso industry by owner and roaster Gong Gyeong-sik (공경식). Starting as a barista at Gangnam chain Coffee Works in 2010, Gong went on to establish his own brand Reissue in 2015. Initially using imported Italian brand Musetti Rossa beans, Reissue currently roast their entire line-up of two blends, four single origins and a decaf in-house on their trusty Diedrich IR-5 infrared roaster. For added variation, Reissue also offers a selection of three different sugars including organic, La Perruche pure cane sugar and their signature vanilla sugar. We recommend going for a Medium Blend Espresso, spiced up with a teaspoon of vanilla sugar for a perfectly balanced espresso.

Coffee Beans As well as two house blends, Medium and Dark, Reissue Coffee sell all their single origins in store, available in both 200g and 1kg bags.

Lauter Coffee

라우터커피

📍 12 Yeonhui-ro 11ma-gil (Yeonhui-dong), Seodaemun-gu, Seoul
🚇 Hongik University (Line 2, Airport Line, Gyeongui-Jungang Line, Exit 3)
📷 @lautercoffee   🏪 **Mon to Sun** 8:00am - 8:00pm

Established **2020**   Brewing Method **Espresso, Filter (V60), Cold Brew**
Recommended Menu **Single Origin Filter, Caffé Latte**

Behind the Counter:
Machine **La Marzocco Linea 2 Group**   Grinder **Mahlkönig EK43, Peak**
Roaster **Probat Probatone 5, Union Hand Roaster**

Head through Yeonnam-dong under the Yeonhui bypass (연희 IC) and you will find one of Seoul's hidden gem cafe suburbs. Quietly tucked away in the side streets and slopes of Yeonhui-dong, the area boasts dozens of cafes, quaint bakeries and unique fusion eateries. Perhaps the most local of them all, Lauter Coffee, opening daily at 8am, serves up smooth lattes and perfect filter brews to queues of regulars who line the steps for their daily caffeine fix.

Combining several years experience working for Hell Cafe Roasters and Deep Blue Lake, head roaster and owner Lee Chang-hoon (이창훈) opened Lauter Coffee in the midst of Covid-19, in October 2020. Popular for its casual setting, friendly service and endless selection of filter options, business at this converted corner shop is going strong as it continues to cement itself at the heart of the community.

Considering the relatively small size of the cafe, Lauter Coffee has an extremely diverse line-up of specialty coffee beans. Off-site their roasting lab churns out House Blend "Trinity No. 1" and single origins for B2B distribution, while in the back of the

cafe Lee produces micro batches of premium beans on a Japanese Union Hand Roaster. Capable of roasting batches as small as 300g, this is ideal for handling their staggering line-up of more than 20 single origins. Given the high calibre and sheer cost of these coffees (think Panama Ninety Plus Juliette, Panama Finca Deborah and Elida Estate "Torre Lot" Geisha) it is imperative to keep them as fresh as possible by roasting only on demand.

Wall Art Note the hundreds of creative memos and post-it notes stuck to the wall by Lauter fans, local artists and fellow coffee professionals.

Homely Vibe Lauter Coffee's casual vibe and friendly team of baristas are sure to make you feel instantly welcome. It may be slightly off the beaten track but don't be put off; this is a must-visit cafe for those looking to get as close to the local coffee culture as possible.

Identity Coffee Lab

아이덴티티커피랩

📍 390 Moraenae-ro (Yeonhui-dong), Seodaemun-gu, Seoul   🚇 Hongje (Line 3, Exit 3)

📷 @identity_coffeelab   🏪 **Thur to Tue** 11:00pm – 7:30pm / **Wed** Closed

Established **2018**   Brewing Method **Espresso, Filter (V60)**
Recommended Menu **Single Origin Filter Coffee**

Behind the Counter:
Machine **Synesso MVP Hydra 3 Group**
Grinder **Victoria Arduino Mythos One, Mahlkönig EK43, Mazzer Robur**
Roaster **Probat, Giesen W6, ROEST S100**

Starting out as a micro-roastery and distributor in 2018, Identity Coffee Lab has quickly become a key player in the industry, steadily expanding their presence both offline and online over the last four years. The company now boasts a showroom cafe in Seoul's Yeonhui-dong (연희동) while roasting takes place off-site in

their newly opened roasting lab. Looking for a fresh start, couple Yoon Won-gyun (윤원균) and Yeom Seon-young (염선영) both quit their jobs to follow their dream of setting up an independent cafe and have never looked back. Yoon heads up the roasting side of the business while Yeom takes charge of the brand's marketing commitments and barista training program.

Located in a quaint residential area north of Yeonhui-dong (연희동), Identity Coffee Lab has a relaxed living-room style interior complete with comfy leather sofas, mahogany furniture and loads of natural light flooding in through floor-to-ceiling street-facing windows. Blending in seamlessly to the background, Identity Coffee Lab has an all white flawless set-up of coffee making machinery including Synesso MVP Hydra 3 Group espresso machine, Victoria Arduino and Mahlkönig grinders.

After first opening up as a drip coffee bar only, the couple-run brand has developed a strong reputation for satisfying locals with lattes and signature espresso variations as well. Using house blends "Chillin" (Guatemala, Brazil, Burundi) and "Mid-Century" (Ethiopia Natural 60% and Ethiopia Washed 40%), check out options like Coffee Shake, White Oak and Vanilla Latte to get a taste of something extraordinary.

When it comes to the brew bar, choose from a selection of different filter beans. Past highlights include everything from classic Ethiopia Worka, Cup of Excellence (CoE) Guatemalas and world-class Geisha from Colombia and Panama. In order to guarantee consistent roasting profiles, Identity Coffee Lab uses a combination of high end machinery from Probat (Germany), Giesen (Netherlands) and ROEST (Norway).

Huelgo

후엘고

📍 118 Mapo-daero 11-gil (Yeomni-dong), Mapo-gu, Seoul
🚇 Daeheung (Line 6, Exit 2)   📷 @huelgocoffee
📅 **Mon to Sat** 11:00am - 8:00pm / **Sun** 11:00am - 6:00pm

Established **2018**   Brewing Method **Espresso, Filter (V60), Cold Brew**
Recommended Menu **Brewing Coffee / Small Latte**

Behind the Counter:
Machine **La Marzocco Linea 2 Group**
Grinder **Victoria Arduino Mythos One, Mahlkönig EK43, Guatemala**   Roaster **Probat**

Tucked away at the top of the hill, Huelgo has been satisfying locals with fresh hand-drip coffee and homemade baking for four years now. Located around 15 minutes walk from the main Mapo-daero (마포대로) thoroughfare, this cafe is testament to specialty coffee's growth at a neighbourhood level, a "third place" for residents to interact at the heart of the community.

Almost the entire street-facing facade of Huelgo is lined with wall-length windows, filling the space with plenty of natural light and offering views out onto the peaceful suburb of Yeomni-dong (염리동). If you miss out on the popular window-counter seats, the cafe has a range of seating options around the bar and a central dining table ideal for studying or group

catch-ups.

Roasting off-site at their Mokdong (목동) roastery, Huelgo's line-up of beans includes decaf single origins, rotating seasonal blends and a selection of Geisha from Colombia, Peru and Costa Rica. Priced around 8,000 Won these might be some of the most affordable Geisha in the city! Pair one with a butter biscuit, pistachio financier or chocolate chip cookie and enjoy the quiet surroundings.

The easiest way to approach Huelgo is by exit 2 of Daeheung station, turning left at the sloping Baekbeom-ro 25 gil (백범로25길) running parallel with the Seoul Design High School (서울 디자인 고등학교). At the peak of the hill look out for the red brick building with Huelgo occupying the ground floor.

YM Espresso Room

YM 에스프레소 룸

📍 Ground Floor, Raemian 909-dong, 43-9 Jingwan 3-ro (Jingwan-dong), Eunpyeong-gu, Seoul
🚇 Gupabal (Line 3, Exit 1)　📷 @ymespressoroom　🏪 **Mon to Sun** 9:00am - 10:00pm

Established **2020**　Brewing Method **Espresso**
Recommended Menu **Espresso / Caffè Latte**

Behind the Counter:
Machine　**La Marzocco Linea 3 Group, Zeroth Law Real 9**
Grinder　**Mazzer ZM, Robur, Victoria Arduino Mythos One, Mythos Two, Anfim SP II**
Roaster　**Fuji Royal 5kg, Diedrich DR-25**

On the northern perimeter of Seoul, in suburban Gupabal, sits one of the most eccentric cafe settings in the entire city. Inspired while working in London and Budapest, YM brand director Cho Yong-min (조용민) has transformed the ground floor of an apartment complex into a mock-up European cathedral, complete with stained-glass windows, service altar and rows of wooden pews. The attention to detail and quality of design, paired with impeccable service make this a must-visit stop on any Seoul cafe tour. The YM Coffee Project comprises three different venues, YM Coffee Project, YM Espresso Room and the recently constructed YM Roasting Factory. Both cafe spaces are brimming with character and leave an equally deep impression on the customer. The former has been running now for over 5 years while their most recent venture, the Espresso Room opened in the midst of Covid-19 during the summer of 2020.

Aside from running two cafes YM Coffee Project roasts coffee beans for wholesale, runs barista

training courses and cafe consultant services. Coffee enthusiasts are welcome to join their coffee bean subscription service or pay for YM's "Home Cafe Classes" where each participant can take a crash course in how to brew their own pour over coffee.

Line-Up Atop the altar sits their prized La Marzocco Linea 3 Group Espresso machine, supported by Mazzer, Victoria Arduino and Anfim grinders.

YM Coffee House YM Coffee House (YM 커피하우스) brings a whole new concept of "living room" interior to the cafe industry. Set in a European style sitting room, each seat along the wooden benchtop faces the central

serving area, bringing you as close to the brewing as possible.

Coffee Cups The coffee cups are all imported or bought directly from Europe including rare Wedgwood (England) and Ginori (Italy) pieces as well as hand-painted German, Russian and Turkish sets too.

Alternative Worship YM Espresso Room's cathedral interior makes for one of the most unique espresso settings imaginable.

Other Branch:
• **YM Coffee House**
  📍 21-8 Yeonseo-ro 29-gil (Galhyeon-dong), Eunpyeong-gu, Seoul
  🚇 Yeonsinnae (Line 3 and 6, Exit 7)

Summit Culture

써밋 컬쳐

📍 11 Sinchon-ro 14 an-gil (Nogosan-dong), Mapo-gu, Seoul
🚇 Sinchon (Line 2, Exit 8)   📷 @summit.culture
📅 **Mon to Sun** 11:00am - 8:00pm

Established **2018**   Brewing Method **Espresso, Filter (V60), Cold Brew**
Recommended Menu **Special Single Origin (Filter)**

Behind the Counter:
Machine **Dalla Corte XT 2 Group**   Grinder **Mahlkönig EK43, Mazzer Robur, Anfim SP II**
Roaster **Easyster 4kg**

With a decade of experience working at Coffee Temple, Shin Jong-cheol (신종철) eventually opened his own venture, Summit Culture, in 2018. Hidden down the backstreets of Seoul's biggest student neighbourhood, Summit's compact space is a popular hangout for locals looking for in-house roasted coffee and freshly baked desserts.
Using a local brand Easyster roaster, Summit's line-up includes a diverse

range of seasonal single origins, two house blends (Sherpa and Grand Bleu) and specialty auction lots under their "Special Single Origins" menu. Summit is run concurrently with patisserie brand Dough (도우), check out the counter top for delicious baked goods like their caramel croffle and Victoria cake or try pairing your filter brew with a crunchy mocha flavoured biscotti.

To find Summit Culture, come out of exit 8 of Sinchon subway station, opposite the Hyundai Department Store, and head down the second alleyway Sinchon-ro 16-gil (신촌로 16길). After a short walk, turn right at the end of the road and the cafe is located on the ground floor of the marble-clad office building. For those keen on a short stroll after coffee, simply walk one more block and you will find yourself on one of the city's most popular trails, the Gyeongui Line Forest Park (경의선 숲길 공원), a 6km urban walkway stretching between Yeonnam-dong (연남동) in the west and Hyochang-dong (효창동) in the east.

Biroso Coffee

비로소 커피

📍 42 Gwangseong-ro 6-gil (Sinsu-dong), Mapo-gu, Seoul

🚇 Daeheung (Line 6, Exit 4)　　📷 @birosocoffee　　🕐 **Mon to Sun** 10:00am - 10:00pm

Established **2016**　　Brewing Method **Espresso, Filter (V60), Cold Brew**
Recommended Menu **Single Origin Filter Coffee**

Behind the Counter:
Machine　La Marzocco Linea 2 Group
Grinder　Mahlkönig E65S, EK43, Mazzer Robur, Ditting KR804
Roaster　Probat UG15, Stronghold

One of the first specialty roasters to settle on the Gyeongui Line Forest Park (경의선 숲길 공원) walkway, Biroso Coffee has served this popular neighbourhood for over six years. Not far from Summit Culture (써밋 컬쳐) this is another great option to add on any coffee excursion in Seoul's Mapo-gu (마포구) suburb. Occupying a bright red brick building, just a few metres off the track, this charming two storey cafe includes a ground floor roasting lab, sleek coffee bar and a spacious second floor with views out onto the gardens below.

Before heading upstairs, choose from a full coffee menu of filter and espresso options as well as Biroso specials, from seasonal ades and Einspänner's, to pair with a selection of tasty desserts. Like a growing number of specialty cafes in Korea, Biroso Coffee places an emphasis on controlling every aspect of the coffee making process in-house, from roasting, quality control and cupping to pulling the same great tasting espresso shots across their team of trained baristas. Biroso's roasting lab is capable of handling up to ten different single origins and two house blends at any one time, carefully roasting each batch on a duo of vintage cast iron Probat UG series and Stronghold roasters.

Coffee and cake As well as roasting all their coffee-in house, Biroso's first floor counter is lined with classic European-influenced recipes from carrot cake, pound cake, financiers and crowd pleasing tiramisu.

Fritz *

프릳츠

🔲 17 Saechang-ro 2-gil (Dohwa-dong), Mapo-gu, Seoul

🔲 Mapo (Line 5, Exit 3), Gongdeok (Line 5 and 6, Airport Line, Exit 8, Exit 9)

🔲 @fritzcoffeecompany　　🔲 **Mon to Fri** 8:00am - 10:00pm / **Sat to Sun** 10:00am - 10:00pm

Established **2014**　　Brewing Method **Espresso, Filter (V60)**

Recommended Menu **Americano**

Behind the Counter:

Machine **Slayer Espresso 3 Group**　　Grinder **Mahlkönig EK43, K30 Twin 2.0**

Roaster **Giesen**

Perhaps the most powerful brand in Korea's specialty coffee scene, Fritz has been setting trends in the industry for the last eight years since its establishment in 2014. It's no surprise the founding crew represent some of the most talented people on the scene; an A-list of industry celebrities that includes Heo Min-su (pastry chef), Kim Byeong-ki (coffee importer and former employee of Coffee Libre), Kim Do-hyun (coffee roaster) and Park Geun-ha (barista and national champion).

At its core Fritz is a coffee importer and roaster, selling their direct-import beans in store, through partner cafes across the country and via their coffee subscription service, the "Fritz Coffee Club".

Their team of bakers are just as serious though, using only French AOP and Danish butter, organic French wheat flour and organic ingredients for every item on the menu. Look out for the bakery timetable at the entrance with information on the day's baking schedule (main photo).

Fritz Core Members Now employing

over 70 people, the brand currently includes five industry experts.

Kim Byeong-ki ("BK") (김병기) - Green Bean Buyer, CEO and Brand Director

Park Geun-ha (박근하) - Roaster, Green Bean Buyer and CEO

Kim Do-hyun (김도현) - Coffee Roaster

Song Seong-man (송성만) - Barista

Jeon Gyeong-mi (전경미) - Coffee Judge

Coffee Beans Fritz Coffee trade directly with around 10 different farms, importing up to 30 different coffees at any one time. As well as a diverse range of single origin beans, Fritz also distribute and sell in-store three different blends; Seoul Cinema (Ethiopia 70%, Costa Rica 30%), Old Dog (Costa Rica 35%, El Salvador 25%, India 40%) and Everything Good (잘 되어가시나) (Ethiopia 25%, Guatemala 20%, Costa Rica 35%, India 20%). A great deal of the coffees come from the same farms used as Blue Bottle, Stumptown and Intelligentsia Coffee. As of 2022 Fritz coffees are also available in capsule and drip bag format too.

Goods From clothing, badges and coffee cups to posters and eco bags, the in-house design team are constantly on-point with their merchandise. If you're looking for a souvenir or present from a Korean cafe, Fritz will not disappoint!

The seal of approval Spot "Fritz" (프릳츠 in Korean), the coffee drinking mascot, on almost everything from the cups, posters and packaging to their custom designed espresso machine (Kees van der Westen Spirit Triplette), badges, cups and gift sets.

The genius of Fritz Coffee's "newtro" (new retro) concept Fritz has been incredibly successful in creating a fusion of foreign cafe culture with local Korean vintage. Gone are the Campbell Soup tins and other classic symbols of nostalgia, Fritz is putting its own mark on the term retro by re-interpreting the 1970s and 1980s with a strong local twist.

Other Branches:

- **Fritz (Yangjae - 양재):**
  - 📍 24-11 Gangnam-daero 37-gil (Seocho-dong), Seocho-gu, Seoul
  - 🚇 Yangjae (Line 3, Exit 1, Exit 2)
- **Fritz (Wonseo - 원서):**
  - 📍 83 Yulgok-ro (Wonseo-dong), Jongno-gu, Seoul (Located on the ground floor of the Arario Museum)
  - 🚇 Anguk (Line 3, Exit 3)

"Coffee and Bread Make Everything Better"

Young and Daughters

영앤도터스

📍 Level 1 (107), Prugio City, 156 Mapo-daero (Gongdeok-dong), Mapo-gu, Seoul
🚇 Gongdeok (Line 5 and 6, Airport Line, Exit 4)   📷 @younganddaughters
🗓️ **Mon to Fri** 7:00am - 5:00pm / **Sat to Sun** 10:00am - 5:00pm

Established **2020**   Brewing Method **Espresso, Batch Brew, Filter (V60)**
Recommended Menu **Seasonal Menu**

Behind the Counter:
Machine **La Marzocco Linea 2 Group**   Grinder **Mazzer Robur**
Roaster **Diedrich IR-5**

Another example of Seoul's changing coffee culture, Young and Daughters opens everyday (midweek) from 7am, serving fresh batch brew (pour over using machine) to early risers on Mapo-gu's busiest high street.

Branding is on-point with 70s-80s themed Korean "newtro" highlights throughout, centred around the brand's Jelly Baby inspired mascots Natalie and Collin.

Seasonal Specials feature homemade caramel recipes paired with in-house roasted coffee including Melting Latte, Deep Caramel Shake (Summer) and Deep Caramel Latte (Winter).

Gangnam

Lull Coffee

룰커피

◎ 16 Banpo-daero 7-gil (Seocho-dong), Seocho-gu, Seoul
🚌 Nambu Bus Terminal (Line 3, Exit 5)　　◎ @lullcoffee_
🗓 **Tue to Sun** 10:00am - 6:00pm / **Mon** Closed

Established　**2020**　　Brewing Method　**Espresso, Filter (Kalita, December Dripper, Steadfast)**
Recommended Menu　**Brewing Coffee**

Behind the Counter:
Machine　**Synesso MVP Hydra 2 Group**
Grinder　**Mahlkönig EK43, Victoria Arduino Mythos One**　　Roaster　**Easyster 1.8kg**

Specialty hand-drip coffee house Lull Coffee, short for "Love You All", is one of our favourite spots on Gangnam's thriving coffee scene. Located a short walk from Nambu Bus Terminal and the impressive Seoul Arts Centre (예술의전당), this inviting, minimalist space is a massive hit with Seoul coffee aficionados and lucky locals.

Head roaster and owner Jeong In-

seong (정인성), a former Korea Brewers Cup Championship (2013) winner, has created a unique bean-to-cup coffee experience combining all of his skills for the perfect pour over. As well as filter brews, Lull also serves espresso based classics, Viennese-style Einspänners, herbal teas and a range of homemade baked goods. Communicating with customers in-between carefully timed pours, Jeong's friendly approach brings a relaxed neighbourhood vibe to this traditionally fast-paced suburb of Gangnam.

Lull roast all of their blends and single origin beans on a small batch Korean-manufactured Easyster machine. Established by entrepreneur Byeon In-kyu (변인규) in 2010, the Easyster brand has become the staple machine of countless cafes and micro-roasters both in Korea and increasingly overseas too, with customers around the world in Australia, Canada and China.

Terarosa*

테라로사

📍 Level 1, Posco Center, 440 Teheran-ro (Daechi-dong), Gangnam-gu, Seoul

🚇 Seolleung (Line 2, Suin-Bundang Line, Exit 1), Samseong (Line 2, Exit 4)

📷 @terarosacoffee   📅 **Mon to Fri** 7:30am - 9:00pm / **Sat to Sun** 8:30am - 9:00pm

Established **2002**   Brewing Method **Espresso, Filter** (Kalita Wave)
Recommended Menu **Today's Drip Coffee**

Behind the Counter:
Machine **Synesso MVP Hydra 3 Group**   Grinder **Ditting KR804, Mazzer Robur**
Roaster **Petroncini Impianti TT8/10, TT15/20, TT60, TT120**

Celebrating their 20th anniversary this year, Terarosa has grown exponentially since opening their first branch in Gangneung. The brand now includes 19 locations across the country including Jeju, Pohang, Busan and Seoul.

Located in the heart of Gangnam's affluent Samseong-dong (삼성동) neighbourhood, Terarosa's huge flagship Seoul cafe is spread across two mezzanine floors of the impressive Posco Center skyscraper. One of the top-ten biggest companies in the country, the steel-manufacturing powerhouse has 18,000 employees and is the fourth largest producer in the world of crude steel.

Complementing the backdrop of the building, Terarosa's interior cleverly incorporates steel throughout the design, with galvanised steel benches, oversized metal staircases and plenty of exposed steel beams, industrial nuts and bolts. With loads of different seating options across each floor, this is a great place to spend a few hours studying, relaxing with friends or simply taking a short break from the hustle and bustle of Gangnam.

Terarosa's menu focuses primarily on hand drip coffees with up to eight different single origins available at

any one time, each sourced directly through Terarosa's in-house green bean and quality control team. Look out for the "Today's Drip" discount marker on the menu, with both a hot and an ice option available for hand drip filter coffee. With an assortment of baked goods available too, make sure not to miss out on their croissants, ciabattas and famous pecan pie to pair with your brew.

For those interested, the cafe is adjacent to Posco's "Steel Gallery" as well, easily accessible from the top of the stairs on the second floor. Other features of the building include a massive 9m aquarium which rises from the basement floor to the lobby. With more than 30 species of South Pacific coral reefs, 2,000 tropical fish, turtles and eels, it certainly makes for an exotic entrance to this grand building!

Gangneung HQ Occupying its own industrial complex, Terarosa Gangneung features a coffee museum, gift shop, bakery and huge converted warehouse cafe. The coastal city of Gangneung is also home to Park Yi-chu's (박이추) Bohemian Coffee Factory as well as a growing number of well-established independent cafes, earning this popular tourist spot the nickname "coffee city".

© Lee Jong-geun, GuruVisual

502 Coffee Roasters*

502커피로스터스

📍 142 Teheran-ro (Yeoksam-dong), Gangnam-gu, Seoul

🚇 Yeoksam (Line 2, Exit 3)　　📷 @502coffeeroasters

📅 **Mon to Fri** 7:30am – 7:30pm / **Sat to Sun** Closed

Established **2009**　　Brewing Method **Espresso, Filter (V60)**
Recommended Menu **Single Origin Filter**

Behind the Counter:
Machine **Synesso MVP Hydra 3 Group**　　Grinder **Anfim SP II, Mahlkönig EK43**
Roaster **Easyster**

Established by Lee Hyeon-jeong (이현정) and Kim Sam-choong (김삼중) in 2009, 502 Coffee Roasters has come a long way since first opening up their hole-in-the-wall takeaway spot in Seoul's Gasan Digital Complex. The team now operate a substantial roasting and distribution business while trading directly with coffee farms across Central and South America in Colombia, El Salvador, Costa Rica and Honduras.

Interior Re-designed in 2020 by brand curator Studio Stof (스튜디오 스토프), 502 Coffee Roasters' flagship showroom contrasts dramatically with the fast-paced city life outside in Gangnam.

Coffee Time Offering respite to the thousands of nearby office workers, 502 Coffee Roasters serves up a diverse range of specialty single origin brews for filter and espresso, paired with British classics lemon pound cake and scones with Rodda's clotted cream.

Leesar Coffee*

리사르커피

📍 60 Dosan-daero 99-gil (Cheongdam-dong), Gangnam-gu, Seoul
🚇 Apgujeong Rodeo (Suin-Bundang Line, Exit 2) 📷 @leesarcoffee
📅 **Mon to Sat** 11:00am – 7:30pm / **Sun** Closed

Established 2012   Brewing Method Espresso
Recommended Menu Caffè Espresso / Caffè Piena

Behind the Counter:
Machine La San Marco 85 Leva Class 3 Group   Grinder Mazzer Robur
Roaster The San Franciscan Roaster Company SF-25E

Now ten years old, the Leesar Coffee brand is indisputably the original espresso trend-setter in Korea. Starting from a tiny Naples style standing bar in February 2012, brand director and CEO Lee Min-seob (이민섭) has since created a roastery and espresso bar cafe near Yaksu station (약수역), followed by a bold venture into Seoul's wealthy suburb of Cheongdam-dong. Their newest cafe has a distinct European vibe to it; luxurious marble countertops, vintage wooden terrace-style chairs and a gold leaf typeface logo, "Better Than Espresso", on the front window. Even the retro lamppost feature looming over the back bar was imported directly from Paris. Don't expect to find any Lavazza or Kimbo beans here; all of Leesar's coffees are roasted in-house at their Yaksu-dong branch. Their current house dark blend features both Brazilian and Ethiopian coffee beans, picked specifically to suit their style of sweet, nutty and intense hand-pulled espressos. The menu has recently expanded to include a full range of seven different white and black espresso variations including traditional takes on Italian classics Piano, con

Panna, Oneroso and Affogato. Building on his experience travelling in Naples, Lee also serves up Caffè Strapazzato, a direct nod to the legendary drink from Naples' Gran Caffè Gambrinus. Sticking to tradition, Leesar baristas use a 3-group La San Marco 85 Leva Class espresso machine which uses a lever style operation to manually force the water through the coffee as opposed to steam pressure extraction in regular machines.

On a busy day, Leesar serves over 400 espressos! If you want to beat the queue try making it to their Yaksu marketplace branch for first order at 7am, the perfect start to the day for any caffeine-seeking early riser.

Posters Leesar's branded posters are a parody on the classic 1920s art by Leonetto Cappiello, "the father of modern advertising" who famously produced a series of adverts for Italian espresso machine manufacturer La Victoria Arduino in 1922.

Other Branches:

• **Leesar Coffee (Yaksu - 약수)**

 16-7 Dasan-ro 8-gil (Sindang-dong), Jung-gu, Seoul

 Yaksu (Line 3, Exit 5 or Line 6, Exit 7)

Gray Gristmill

그레이 그리스트밀

📍 15 Apgujeong-ro 2-gil (Sinsa-dong), Gangnam-gu, Seoul
🚇 Sinsa (Line 3, Exit 6)　　📷 @gray_gristmill
🗓 **Mon to Sun** 11:00am – 9:00pm

Established **2018**　　Brewing Method **Espresso, Filter (V60)**
Recommended Menu **Specialty Single Origin Filter Coffee**

Behind the Counter:
Machine **Slayer Espresso 1 Group**　　Grinder **Mahlkönig EK43**
Roaster **Loring S15 Falcon**

This is coffee, Gangnam style. Right in the heart of the city's fashion district, Sinsa-dong (신사동), watch the team of Seoul's best-dressed baristas put on their finest performance at Gray Gristmill's (그레이 그리스트밀) specially arranged brew bars. Designed to replicate the environment of the competitive barista circuit, put yourself in the shoes of the judging panel and be prepared for a refined coffee experience like none other.

This state-of-the-art roastery cafe was established in 2018 by two champion baristas Kim Hyeon-soo (김현수) and Bang Jun-bae (방준배), along with

support from Gangnam-based cafe consultancy Andrea Plus (안드레아 플러스).

Coffee sommelier Each of their ten different single origin coffee beans are packed neatly in single portion vacuum packs, ready to serve each customer one-by-one. Simply pick your style of beans at the counter then choose your favourite type of coffee; americano, latte, brewing or affogato.

Bang Jun-bae (방준배) Before getting behind the counter at Gray Gristmill,

Bang worked as barista trainer and latte art coach, helping countless numbers of coffee professionals enter the trade and compete on the local barista circuit. Bang himself competed in the Korea National Barista Championship (KNBC) in four consecutive years, winning the event in 2017 and placing second in 2016. After four years he recently left Seoul to start his own cafe in Daegu, joining a small cohort of Gray Gristmill veterans who continue their finely tuned trade under their own brand.

Incheon, Gyeonggi-do and Daejeon

Incheon, Gyeonggi-do and Daejeon

① Incheon
② Seongnam-si   ③ Yongin-si
④ Pyeongtaek-si   ⑤ Daejeon

Bean Brothers 194p

Dohwa

Chromite Coffee 192p

Developing Room 190p

Songdo

Yeonsu

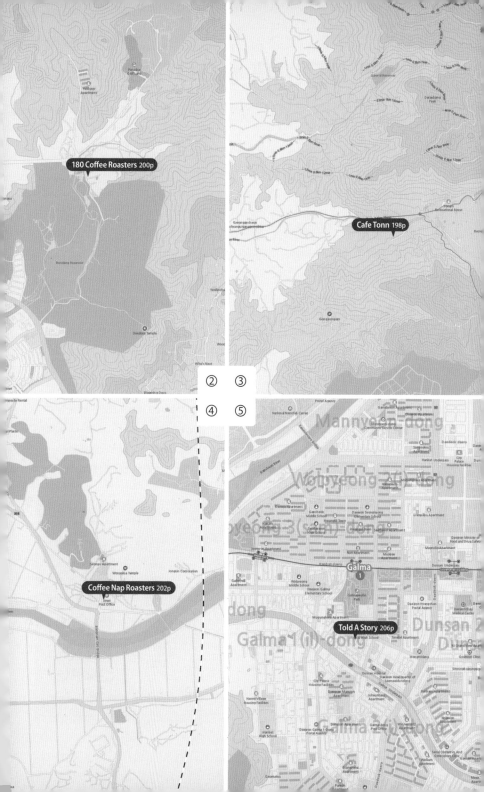

180 Coffee Roasters 200p

Cafe Tonn 198p

②　③

④　⑤

Coffee Nap Roasters 202p

Told A Story 206p

Developing Room

디벨로핑룸

📍 26 Cheongmyeong-ro (Cheonghak-dong), Yeonsu-gu, Incheon
🚇 Yeonsu (Suin-Bundang Line, Exit 4)   📷 @developingroom
🏪 **Mon to Sun** 11:00am - 10:00pm

Established **2017**   Brewing Method **Espresso, Filter (V60)**
Recommended Menu **Brewing Coffee**

Behind the Counter:
Machine La Marzocco FB80 2 Group
Grinder **Mahlkönig GH2, EK43, Mazzer Robur, Anfim SP II**   Roaster **Diedrich IR 2.5, IR 5**

Opened in 2017, Developing Room is one of the longest running specialty cafes in the Incheon area. Going strong for five years now, the business was established by local Park Gi-beom (박기범) who returned to his hometown after a long stretch working with Seoul-based coffee brands Paul

Bassett and Anthracite Coffee Roasters (앤트러사이트 커피).

The expansive studio-like space has a strong community feel to it, with large workbench seating areas, regular public cupping and brewing classes and even the occasional live music concert. Behind the laid-back workshop vibe though, Developing Room runs a substantial B2B coffee distribution business, working closely with partner cafes across Incheon and the wider Gyeonggi-do province. Using a pair of USA-manufactured Diedrich roasters (IR 2.5 and IR 5 models), the in-house team offer up an impressive range of

eight different specialty single origins at any one time, along with two house espresso blends Stay and Standard Room.

Developing Time Those familiar with coffee roasting will recognise the term, referring to the time it takes between the "first crack" and the end of the roast.

Wall Art All the photos lining the walls are curated by the owner, a keen photographer and designer, with many of them developed from Park's own collection of film and DSLR cameras.

Chromite Coffee

크로마이트커피

📍 39-5 Cheongnyang-ro 155 beon-gil (Ongnyeon-dong), Yeonsu-gu, Incheon
🚇 Songdo (Suin-Bundang Line, Exit 1)   📷 @chromitecoffee
🕐 **Mon to Sat** 12:00pm - 10:00pm / **Sun** 12:00pm – 9:30pm

Established 2016   Brewing Method Espresso, Filter (V60, Siphon), Cold Brew
Recommended Menu Brewing Coffee / Single Latte

Behind the Counter:
Machine La Marzocco Linea 2 Group
Grinder Mahlkönig K30 Twin, EK43, Lean Weber EG-1
Roaster Probat P12, Fuji Royal 5kg, Easyster 1.8kg, Buja Roaster 550

The successor to Busan-based "Coffee Story" (established 2002), the Chromite Coffee team moved north to Incheon in 2016. Soon re-adjusting to their new surroundings, the brand has grown from strength to strength with regulars and coffee fans from across the Gyeonggi-do province queuing up every weekend for a seat in this cosy coffee house.

Occupying an entire hilltop villa, the space has been cleverley remodelled with a central brew bar and service area, branching off to several rooms each with their own unique atmosphere. Roasting on-site, the brand operates an impressive coffee lab on the basement floor while upstairs industry professionals teach Specialty Coffee Association (SCA)-accredited barista skills courses. Heading up the Korean chapter of the SCA since 2018, Chromite Coffee works in partnership with global experts to offer brewing, roasting and sensory skills classes as well as hosting a number of principal events on the Korea National Barista Championships (KNBC) circuit.

When it comes to ordering coffee,

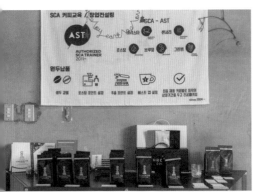

Chromite serves up a menu unlikely to be found anywhere else in Incheon. From in-house roasted Yemen Jaadi and Colombian Pink Bourbon filter options, to single origin lattes, siphon and cold brew, this is a must-visit cafe for anyone stopping by Songdo.

Operation Chromite Named after the code name of the decisive Incheon Landing Operation (인천상륙작전) during the Korean War, Chromite Coffee was one of the first specialty coffee roasters to open in this growing international port city.

Bean Brothers

빈브라더스

📍 12 Sukgol-ro 88 beon-gil (Dohwa-dong), Michuhol-gu, Incheon
🚇 Dohwa (Line 1, Exit 3) 📷 @bean_brothers
📅 **Mon to Sun** 11:00am - 9:00pm

Established 2013 Brewing Method Espresso, Filter (Clever Dripper), Cold Brew
Recommended Menu House Blend Americano

Behind the Counter:
Machine La Marzocco GB5 Grinder Ditting KR804, Mazzer Robur
Roaster Loring S35 Kestrel, Stronghold S7

Roasting coffee since 2013, the Bean Brothers operation includes six cafes in the Gyeonggi-do province, a distribution network of over 200 cafes and a warehouse capable of storing 100 tons of imported coffee beans at any one time.

Alongside daily house espresso blends "Black Suit" and "Velvet White", the brand has a huge following for their recently launched "Auction Series", a premium line-up of extremely rare and exquisite coffees distributed exclusively in Korea by Bean Brothers. Bought directly by the company's green bean team at local plantation auctions, recent examples include Ninety Plus "Tigre" and "Kulé" as well as two highly awarded coffees from Cup of Excellence (CoE) accredited Colombian grower Mikava Santuario.

Bean Brothers has long been on-

point with the popular "home cafe trend", sending out weekly coffee care packages via their "personal coffee guide" subscription service since day one. Run concurrently with their online magazine and education seminars, this is one of the longest running coffee subscription services in the country.

Launching Overseas One of the first brands to expand into the wider South East Asian market, Bean Brothers also operates two branches in Kuala Lumpur, opened in 2015 and 2018. A further third branch is due to open in late 2022.

Warehouse Vibes Located in a former beer warehouse, their lofty Hapjeong branch still maintains some of the original industrial aesthetic, from exposed steel beams, bare bulbs and rough brick walls.

Other Branches:
• **Bean Brothers (Hapjeong** - 합정**)**
  📍 35-1 Tojeong-ro (Hapjeong-dong), Mapo-gu, Seoul
  🚇 Hapjeong (Line 2 and 6, Exit 7)

Cafe Tonn

카페톤

📍 192 Chobu-ro (Chobu-ri), Mohyeon-eup, Cheoin-gu, Yongin-si, Gyeonggi-do
📵 N/A 📷 @tonn__official, @coffea_original 🚆 **Mon to Sun** 10:00am - 8:00pm

Established **2019** Brewing Method **Espresso, Filter (V60), Cold Brew**
Recommended Menu **Premium Latte / Americano**

Behind the Counter:
Machine **Victoria Arduino Black Eagle 3 Group**
Grinder **Victoria Arduino Mythos Two, Mahlkönig Peak**
Roaster **Bühler 120kg RoadMaster, Giesen W45A, W60A, Probat**

Flagship store of long-established brand Coffea Coffee (2007), Cafe Tonn was eventually opened in 2019 as the company's first B2C venture. Located just a short drive south of Seoul, the dramatic steel and glass structure, complete with gallery, expansive terrace and garden water feature is a popular weekend brunch spot.

Global Player As well as operating a nationwide coffee bean roasting and distribution business, Coffea has a total of 16 overseas branches spread across Malaysia and China. Pre-Covid-19, their team of green bean direct traders worked tirelessly building relationships with their network of over 30 different coffee farms in 14 countries. CEO Choi Ji-wook (최지욱) regularly participates in Cup of Excellence (CoE) auctions in South America and was recently appointed to the board of Fairtrade International's Golden Cup panel, the first Korean to achieve this accolade.

180 Coffee Roasters*

180커피로스터스

📍 4 Munjeong-ro 144beon-gil (Yul-dong), Bundang-gu, Seongnam-si, Gyeonggi-do
💾 N/A   📷 @180coffeeroasters_official
🏪 **Mon to Sun** 9:00am - 6:00pm

Established 2013   Brewing Method Espresso, Filter (V60, Siphon, Aeropress), Cold Brew
Recommended Menu Americano / Creams Evans

Behind the Counter:

Machine **Victoria Arduino VA388 Black Eagle 3 Group**
Grinder Victoria Arduino Mythos One, Mazzer Robur, Mahlkönig EKK43
Roaster Probat Probatino, UG15, Easyster 1.8kg, Stronghold S7, Giesen W1, W30

CEO Lee Seung-jin (이승진) first learnt coffee in 2008 under the stewardship of Coffee Libre's Seo Pil-hoon (서필훈). Just five years later he would go on to take 1st place at the first ever Korean Coffee Roasting Championship (KCRC), later earning himself a spot representing Korea at the subsequent World Roasting Championships.

Coffee made by champions Founding member Lee Seung-jin (이승진) has a team of seven working behind the scenes at their Bundang headquarters. With an astonishing ten different roasting machines, 180 Coffee Roasters

"Changing your expectations 180°"

produce a wide range of coffees including several different espresso blends, seasonal single origins and a range of ready-to-drink (RTD) cold brew options.

Trophy Cabinet The long list of achievements by 180 Coffee Roasters' team of coffee professionals includes:

Korea Coffee Roasting Championship (KCRC) 1st Place 2013 and 2nd Place 2019, World Coffee Roasting Championship (WCRC) 5th Place 2013, Korea National Barista Championship (KNBC) 4th Place 2014 and Korea Aeropress Championship (KAC) 1st Place 2017.

Coffee Nap Roasters

커피냅로스터스

📍 35 Bongnam 2-gil (Bongnam-ri), Jinwi-myeon, Pyeongtaek-si, Gyeonggi-do

🚇 N/A   📷 @coffeenap_roasters   🏪 **Mon to Sun** 10:00am - 8:00pm

Established **2017**   Brewing Method **Espresso, Filter (V60), Cold Brew**
Recommended Menu **Flat White / Single Origin Filter**

Behind the Counter:
Machine La Marzocco GB5 3 Group
Grinder Mahlkönig EK43, K30 Twin, Mazzer Robur
Roaster Giesen W6, Probat Sampler

Celebrating their fifth year anniversary in 2022, the Coffee Nap Roasters brand initially started out in the rural setting of Jinwi village in Pyeongtaek. They've since added two more venues, opening their second branch in Seoul's Yeonnam-dong in 2018, with a third cafe recently added on Jeju island. Their "HQ" and warehouse style cafe in Jinwi remains relatively unchanged since the interior was remodelled in 2017. Features include a four metre long sliding front door, exposed steel beams and framework, and a minimalist zen-like outdoor terrace surrounding the building on all sides.

Inside, the central serving counter is cleverly built across the middle of the warehouse floor, with the brew bar section floating carefully over the exposed floor underneath.

Yeonnam-dong Tucked in between alleyways, re-generated parkways and trendy restaurants, Coffee Nap Roasters (NAP) Yeonnam-dong remains one of the most popular cafes in Seoul's trendy suburb of Yeonnam-dong. Aside from the coffee, NAP Yeonnam-dong is famous for its unique abstract interior. Designed by local studio Maoom, the compact 50m² was awarded the Best Interior Design award at the 2019 ABB Leaf Awards held in Berlin. The bold concept features a mound of 7,000 red bricks, wall-to-wall sliding French door terrace and a single skylight designed to cast a different light on the hill depending on the season or time of night or day. At night the coffee bar, positioned behind the hill, lights up to cast moonlight on the cafe floor and surrounding alleyways.

Roasting Nap roast extensively on their in-house Giesen (Netherlands) and typically have a line-up that includes five single origins alongside three blends, named 317 (India 30%, Guatemala 35% and El Salvador 35%), 218 (Ethiopia x 2) and 420 (Kenya 25%, Costa Rica 25%, Colombia 25%, Ethiopia 25%).

Other Branches:
- **Coffee Nap Roasters (Yeonnam-dong - 연남동)**
  - 📍 70 Seongmisan-ro 27-gil (Yeonnam-dong), Mapo-gu, Seoul
  - 🚉 Gajwa (Gyeonui-Jungang Line, Exit 4)
- **Coffee Nap Roasters (Jeju Island - 제주도)**
  - 📍 45 Hagwi 2-gil (Hagwi 1-ri), Aewol-eup, Jeju-si, Jeju-do
  - 🚉 N/A

Told A Story Coffee Roaster

톨드어스토리 커피로스터

📍 31 Galmayeok-ro 25beon-gil (Galma-dong), Seo-gu, Daejeon
🚇 Galma (Line 1, Exit 1)   📷 @toldastory_coffee_roaster
🕐 **Mon to Fri** 11:00am - 8:00pm / **Sat to Sun** 11:00am - 7:00pm

Established **2005**   Brewing Method **Espresso, Filter (V60), Cold Brew**
Recommended Menu   **House Blend "The King" Americano**

Behind the Counter:
Machine **Synesso MVP 3 Group**   Grinder **Mazzer Robur**
Roaster **Loring S35 Kestrel, Easyster, Fuji Royal, Probat**

Daejeon's flagship specialty coffee brand, Told A Story has been roasting coffee in the city since the early 2000s. Based out of the same shopfront for 16 years (2005 until November 2021), they recently moved to new premises across the river in Daejon's Galma-dong.

Carefully curated by head roaster and owner Kim Geon-pyo (김건표), the no-frills coffee menu includes a combination of classic espresso options, several house blends (The King, Joker and Gorilla) as well as rare single origins sold under their "Gold Label" series. Expect to find everything from Tanzanian and Bolivian Geisha to Cup of Excellence (CoE) and other exclusive auction lots.

Long before the relatively recent boom in Korean specialty cafes, Told A Story has been a gathering place for coffee aficionados who come from far and wide to attend cupping classes and roasting seminars as well as try some of Kim's exceptional blends. Not afraid to push the boundaries,

Kim works constantly behind the scenes in the roasting lab, skillfully operating a series of high-grade roasters. Starting with a Probat BUNN, an American tuning model of the original German Probat, Told A Story have since invested in a Fuji Royal, a locally made Easyster 1kg sampler and a medium-large batch Loring S35 Kestrel model.

Busan and Gyeongsang Province

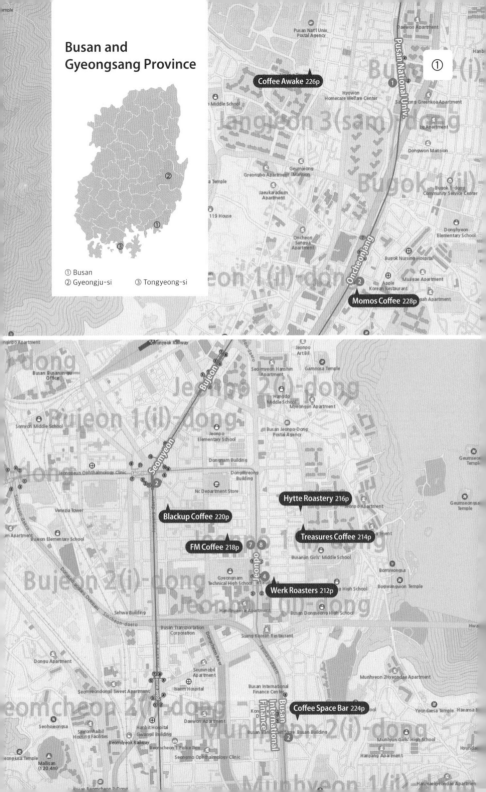

Busan and
Gyeongsang Province

① Busan
② Gyeongju-si
③ Tongyeong-si

①

Coffee Awake 226p

Momos Coffee 228p

Hytte Roastery 216p

Blackup Coffee 220p

Treasures Coffee 214p

FM Coffee 218p

Werk Roasters 212p

Coffee Space Bar 224p

Werk Roasters*

베르크로스터스

📍 115 Seojeon-ro 58beon-gil (Jeonpo-dong), Busanjin-gu, Busan

🚇 Jeonpo (Line 2, Exit 4)   📷 @werk.roasters

📅 **Mon to Fri** 10:00am - 6:30pm / **Sat to Sun** Closed

Established 2017   Brewing Method Espresso, Filter (V60), Cold Brew
Recommended Menu Baby Latte

Behind the Counter:
Machine La Marzocco Linea 2 Group   Grinder Mahlkönig EK43S, Anfim SP II
Roaster Diedrich IR-5, IR-12, CR-35, IKAWA Sampler

Established in 2017, Werk (short for the German word for workshop, "Werkstatt") is one of the biggest names in specialty coffee in the southern city of Busan (부산). Celebrating their fifth anniversary in 2022 they overhauled their flagship branch, moving the service area, drip bar and goods showroom to the first floor, while making slight adjustments to the popular second floor lounge. Lined with plenty of bench seating, laid back tunes, and vertical slit windows this minimalist space is a great place to chill out before embarking on a tour of this exciting neighbourhood.

Werk recently expanded their operation to house their growing range of roasters in a separate off-site workshop. The move signals the start of a new era for the company and an opportunity to expand on their growing range of single origin beans. Their impressive roasting set up includes two infrared Diedrich machines with both a small batch IR-5 (5kg) and medium batch IR-12 (12kg). A third and even larger roaster, the CR-35 was produced specifically for Werk at Diedrich's USA headquarters and proudly installed in Busan last year.

Tasting Set For an all-round tasting experience, go for the Tasting Set. Sit back and enjoy a selection of two drinks from either espresso, americano or latte.

Specialty Beans As well as buying through direct trade, Werk maintains a close relationship with green bean suppliers Namusairo (나무사이로), Coffee Libre (커피 리브레) and Momos Coffee (모모스커피) for a steady supply of premium beans. The range typically includes house blends Baby and Haus alongside a seasonal blend and an assortment of single origins. Find the beans on Werk's online shop, or sample them at one of 100 partner cafes across the country.

Treasures Coffee

트레져스커피

📍 4 Dongseong-ro 49beon-gil (Jeonpo-dong), Busanjin-gu, Busan
🚃 Jeonpo (Line 2, Exit 8)   📷 @treasurescoffee
📅 **Mon to Sun** 12:00pm - 8:00pm / **Wed** Closed

Established **2016**   Brewing Method **Espresso, Filter (V60)**
Recommended Menu **Single Espresso**

Behind the Counter:
Machine **Victoria Arduino VA388 Black Eagle 3 Group**
Grinder **Mahlkönig E80 Supreme, E65, EK43**   Roaster **Stronghold, Easyster 4kg**

A must-visit cafe for anyone travelling to Busan (부산), Treasures Coffee is one of several specialty cafes in the Jeonpo-dong neighbourhood. Other nearby cafes to tick-off on a cafe crawl include Hytte Roastery (just 100 metres away) and Werk Roasters, all accessible via Exit 4 or Exit 8 of Jeonpo station (Line 2).

Originally established in 2016 by roaster Park Joo-hwan (박주환), the business has expanded steadily over

the last 6 years, moving to Jeonpo-dong in 2019, before finally settling at their current converted two-storey villa in February 2021. Choose from a seat at the counter for some front row action at the brew bar or head up the quaint terracotta stairs for some peace and quiet on the second floor.

Treasure Trove Few cafes in Busan can boast a coffee line-up like Treasures Coffee. The menu includes a filter section for Specialty single origin coffees, two house blends (Black Pearl and Ruby) and a rotating Single Specialty Espresso option.

Hytte Roastery

히떼 로스터리

📍 59 Dongseong-ro (Jeonpo-dong), Busanjin-gu, Busan
🚇 Jeonpo (Line 2, Exit 8)   📷 @hytte_roastery
🗓 **Mon to Fri** 12:00pm - 8:00pm / **Thur** closed / **Sat to Sun** 10:00am - 8:00pm

Established **2018**   Brewing Method **Espresso, Filter (V60)**
Recommended Menu **Single Origin Filter Coffee**

Behind the Counter:
Machine **La Marzocco Linea 2 Group**
Grinder **Mahlkönig EK43, E65S**   Roaster **Giesen W6, W15**

Established by couple Jeong Hyo-jae (정효재) and Choi Hee-yun (최희윤), Hytte Roastery is renowned for their North European, or light "Nordic" roasting style. Arguably founded by Oslo-based Tim Wendelboe, the style has a huge following worldwide and is gaining in popularity in Korea thanks to brands like Hytte Roastery, Mesh Coffee (메쉬커피) (Seoul) and Deep Blue Lake (딥블루레이크) (Seoul).

Hytte settled for this style after travelling the world for six months in 2016, ticking off a long bucket list of cafe destinations including Australia, Japan, Denmark and Norway.

Single Origin Batches Fresh single origin batches, using green beans from Namusairo (나무사이로), Coffee Libre (커피 리브레) and Momos Coffee (모모스커피), are roasted in small batches between 1 and 3kg each. Every

week the dedicated team of baristas complete a set of handwritten tasting and brewing notes for customers to review before selecting their brew. For the seasoned aficionado, take a closer look at their public "Calibration Records" and you can read every detail for each roasted bean, from variety, process to roasting level.

Nordic Theme "Hytte" means cottage or log cabin in Norwegian.

FM Coffee*

에프엠커피

📍 26 Jeonpo-daero 199 beon-gil (Jeonpo-dong), Busanjin-gu, Busan

🚇 Jeonpo (Line 2, Exit 7)   📷 @fmcoffee__   🗓 **Mon to Sun** 10:00am - 10:00pm

Established **2010**   Brewing Method **Espresso, Filter (V60), Cold Brew**
Recommended Menu **Today's Coffee / Tomorrow Original** (Cold Brew and Cream)

Behind the Counter:
Machine **Kees Van der Westen Spirit 3 Group**
Grinder **Mahlkönig EKK43, K30, Victoria Arduino Mythos One, Mazzer Robur**
Roaster **Loring S15 Falcon**

Veteran specialty coffee roaster established in 2010 in Busan's Jeonpo-dong (전포동) suburb.
In the last decade this central neighbourhood has become the city's hotspots for trendy cafes with a large concentration of them just a short walk from Jeonpo subway station (Line 2).
FM's fresh line-up of specialty single origin coffees include premier auction lots, Panama Geisha and several coffees sourced directly by roaster Kang Mu-seong (강무성) and his professional green bean sourcing team. Look out for "Today's Coffee" on the menu for their latest specialty offering. For a taste of their house blends, opt for an americano with

either FM Dark, FM Brown and FM Special beans.

Signature Menu Make sure to check out FM's signature menu including seasonal treats like Tomorrow Original (Cold Brew and Signature Cream), Jeonpo-Forest (Matcha Latte and Signature Cream) and Cold Brew Vanilla Latte.

"Field Manual (FM)"

Blackup Coffee*

블랙업커피

📍 41 Seojeon-ro 10 beon-gil (Bujeon-dong), Busanjin-gu, Busan

🚇 Seomyeon (Line 1 and 2, Exit 2)　　📷 @blackup_coffee

📅 **Mon to Sun** 10:00am - 10:00pm

Established **2006**　　Brewing Method **Espresso, Filter (V60), Cold Brew**
Recommended Menu **Single Origin Filter Coffee**

Behind the Counter:

Machine **La Marzocco GS3 1 Group, MOAI Bar System 2 Group**

Grinder **Mahlkönig EK43, Mazzer Robur**　　Roaster **Giesen W15**

Established in 2006 by roaster Kim Myeong-sik (김명식), Blackup Coffee has been at the heart of Busan's specialty coffee scene for over 15 years. Starting in Seomyeon (서면), the brand now includes ten branches and a burgeoning team of over 70 baristas, making it one of the largest specialty coffee brands not just in Busan, but the whole of Korea.

The original branch, set over three floors, makes for the ideal pit stop to

sit back with a fresh filter brew before setting out again into this sprawling coastal metropolis. Choose from a seat at the first floor counter bar, or pick up a slice of homemade cheesecake and head to the ample lounge areas on floors two and three.

For the hand drip connoisseur, Blackup Coffee will not disappoint with a diverse seasonal single origin line-up that few other cafes in the city can match. From floral Panama Geisha to intense Ethiopian Uraga Raros, the line-up typically includes a selection of eight different beans and one decaf option too.

Stock Up Take a look on the first floor for some take-home treats too including capsule coffee, handy coffee drip bags, and ready-to-drink cold brew.

Coffee Space Bar

커피스페이스바

📍 95 Hwangnyeong-daero 74beon-gil (Munhyeon-dong), Nam-gu, Busan
🚇 Busan International Finance Center (Line 2, Exit 2)
📷 @coffee_space_bar
📅 **Mon to Fri** 11:30am - 3:00pm / **Sat to Sun** Closed

Established **2017**   Brewing Method **Espresso**
Recommended Menu **Latte**

Behind the Counter:
Machine **La Marzocco Linea 2 Group**   Grinder **Anfim SP II**

Opened by Busan local Kim Tae-wan (김태완) in 2017, Coffee Space Bar (or CSB) plays a pivotal role in introducing regional coffee roasters Werk Roasters (베르크로스터스), In & Bean Coffee Roasters (인앤빈 커피로스터스) and Hytte Roastery (히떼 로스터리). Literally resembling the space bar on a keyboard, this unique cafe is a hidden gem just a stone's throw away from Busan's Central Business District, Munhyeon-dong (문현동).

Coffee Space Bar opened up a second branch "CSB Lounge" in May 2020, this time collaborating with contemporary artists and photographers as well, curating an interesting coffee drinking experience in Busan's Namcheon-dong (남천동). Set on the ground floor of an unassuming commercial building, this curious space is full of surprises from its secluded tropical terrace garden, sleek floor-to-ceiling windows and after-hours cocktail bar menu. Open till 11:00pm, why not hang around and enjoy a Busan-inspired "Beachside" or choose from an extensive list of gin and tonics featuring top shelf brands The Botanist, Sipsmith London Dry and Bel Air by Distillerie de Paris.

224

Coffee Awake

커피어웨이크

📍 46-4 Busandaehak-ro 63beon-gil (Jangjeon-dong), Geumjeong-gu, Busan
🚇 Pusan National University (Line 1, Exit 1)　　📷 @coffeeawake
🕐 **Mon to Sun** 10:00am - 7:00pm

Established **2012**　　Brewing Method **Espresso, Filter** (Hario Switch), **Cold Brew**
Recommended Menu **Single Origin Filter Coffee**

Behind the Counter:
Machine **Faema E71 3 Group**
Grinder **Mahlkönig EK43, Anfim SP II, Compak PK100, Victoria Arduino Mythos One**
Roaster **Giesen W1, W15**

In the bustling back-alleys of Pusan National University, owner Kim Ji-yong (김지용) has been brewing coffee for locals and students since 2012. Located just 200m from the university's front gate, regulars are drawn in by Coffee Awake's wide range of seasonal single origins, smooth flat whites and

affordable espressos.

Renovated in 2019, the bright street-facing interior has a sleek modern vibe with white marble countertops, mahogany benches and wall-length windows flooding the cosy space with natural light.

Roasting in-house on a pair of trusty Giesen W1 and W15 roasters, Kim handles an impressive five small batch single origins for hand brew and two house blends at any one time.

House Blends "Awaken" (Guatemala 50%, Ethiopia 50%) and "Ordinary" (India 50%, Colombia 30%, Guatemala 20%)

Momos Coffee*

모모스커피

📍 18-1 Osige-ro (Bugok-dong), Geumjeong-gu, Busan
🚇 Oncheonjang (Line 1, Exit 2)   📷 @momos_coffee
🗓 **Mon to Sun** 8:00am - 6:00pm

Established **2007**   Brewing Method **Espresso, Filter (V60)**
Recommended Menu **Today's Hand Drip**

Behind the Counter:
Machine **Victoria Arduino Black Eagle 3 Group**
Grinder **Mazzer Robur, Mahlkönig K30, K43**   Roaster **Probat P25, P60, UG20**

From humble beginnings almost 15 years ago, Momos Coffee has grown into one of the biggest specialty coffee brands in the country. Established in 2007 by Lee Hyeon-gi (이현기) with just 13m² of floor space, the operation now includes a barista academy, over 50 staff, a two storey bakery-cafe and a nationwide coffee distribution business. To top it off the barista team is led by 2019 World Barista Champion (WBC) and Busan local Jeon Joo-yeon (전주연). Roasted beans from Momos Coffee are some of the most highly prized in the market. Roasting on a mighty Probat P 25, the roasting team produce seasonal blends, single origins and actively take part in all of the world's premier auctions including Best of Panama, Cup of Excellence (CoE) and the QIMA Yemen Coffee auction. Following Jeon Joo-yeon's success in the WBC an additional range was also added under the name "Joo Yeon Selection".

If you can't make your mind up

when it comes to the menu, check out the chalkboard at the entrance gate for their "coffee of the day" recommendation.

Layout Dubbed the "Momos Village", the sprawling cafe complex contains on-site bakery, roastery, barista-training school and plenty of different seating options spread across two floors. In an otherwise mundane Korean residential suburb, Momos Coffee appears like a backyard oasis in a concrete jungle, complete with water features, a mini bamboo forest and an array of stone garden ornaments.

Momos Roastery and Coffee Bar The Momos Coffee brand now includes an additional cafe (opened 2021) in Yeongdo (영도), the harbourside suburb of Bongnae-dong (봉래동); built inside an existing warehouse (pictured right), head along to see their state-of-the-art bean storehouse, roasting production line and minimalist brew bar.

Other Branches:

• **Momos Roastery and Coffee Bar**

🔲 160 Bongnaenaru-ro (Bongnae-dong 2-ga), Yeongdo-gu, Busan

🚇 Nampo-dong (Line 1, Exit 8)

Coffee Front

커피프론트

📍 Centum Leaders Mark, 17 APEC-ro (U-dong), Haeundae-gu, Busan
🚇 Centum City (Line 2, Exit 11)   📷 @coffeefront_busan
📅 **Mon to Fri** 8:00am - 7:00pm / **Sat to Sun** 9:00am - 7:00pm

Established **2020**   Brewing Method **Espresso, Filter** (Hario Switch, Chemex)
Recommended Menu C-**Flat** (Signature Flat White with a sugar cane base)

Behind the Counter:
Machine La Marzocco Linea 2 Group
Grinder Mahlkönig EK43, E65S

Inspired by eight years living in San Francisco, owner and seasoned entrepreneur Gu Min-wook (구민욱) combines a wealth of experience in roasting, sensory and barista skills for his latest venture in Busan's vibrant commercial hub, Centum City.
Set discreetly on the ground floor of one of the area's signature skyscrapers,

the cafe boasts an overseas vibe with a central standing espresso bar, ideal to accommodate 8am "rush-hour" coffee crowds. In an area typically dominated by chain coffee shops, Coffee Front is changing the daily routine of local residents and office workers with its trendy menu of specialty blend espressos, signature lattes and a distinctive recording studio style interior.

Using a simplified menu format, coffee is divided here into black (espresso and filter) and white (latte, flat white) options, similar in style to the coffee menus seen in Australia and New

Zealand cafes. When it comes to milk options, Coffee Front have all the bases covered with three different choices available; regular and low fat organic milk by premium label Bumsan Dairy Farm (범산목장) as well as the increasingly popular Oatly plant-based alternative milk.

Also run by Gu Min-wook, Strut Coffee (스트럿커피) (established in 2016) supply all the beans for Coffee Front from their roasting lab and cafe in the western suburb of Gimhae (김해). Using a pair of Giesen (Netherlands) and Diedrich (USA) roasters, the brand runs a successful OEM small batch coffee bean distribution business for a growing number of specialty cafes in Korea's second largest city.

Busan Sightseeing Located just a short drive away from both of Busan's main beaches, Haeundae (해운대 해수욕장) and Gwangalli (광안리 해수욕장), as well as the Busan Exhibition and Convention Center (BEXCO), Coffee Front is the ideal place to explore the city from.

Coffee Place*

커피플레이스

📍 18 Jungang-ro (Nodong-dong), Gyeongju-si, Gyeongsangbuk-do   📖 N/A

📷 @coffeeplace.go   🕐 **Mon to Sun** 9:00am - 7:00pm

Established **2010**   Brewing Method **Espresso, Filter (V60)**

Recommended Menu **Today's Coffee (Filter Coffee)**

Behind the Counter:

Machine **Nuova Simonelli Aurelia T3**

Grinder **Mahlkönig EK43, K30 Twin**

Roaster **Taehwan Proastar THCR-12**

Few buildings in Korea, let alone cafes, have a view like Coffee Place in Gyeongju. Unchanged for centuries, the coffee bar and take-away bench look directly out onto giant Bonghwangdae (봉황대) burial hills from the Three Kingdom Period (삼국 시대), close to 1,500 years old. It's no surprise this historic city, famed for its UNESCO World Heritage tombs and laid back vibe, is known in Korea as the "Museum without walls".

Coffee Place opened in 2010 and has long held the position of

Gyeongju's first and foremost specialty cafe brand. Established in central Nodong-dong (노동동) the brand has gradually expanded its portfolio to include a total of seven cafes across the Gyeongsangbuk-do province, spread evenly throughout Pohang and Gyeongju. Keeping a close eye on changing trends and increasing local competition, the director Jeong Dong-wook (정 동욱) recently completed major renovations to the cafe as it enters its second decade of operation. The interior was completely refitted, new appointments were made to the roasting team and a filter-coffee focused menu was introduced to match the growing demand by regulars and visitors alike to try premium hand-drip coffees.

Now just two hours away from Seoul by bullet train (KTX) this is a must-visit heritage site on any trip to Korea. Head for Coffee Place, pick up a fresh brew to go, and soak up Silla (신라) Dynasty history while walking the sights of this remarkable city.

Sammundang Coffee Company

삼문당커피컴퍼니

📍 168 Jungang-ro (Taepyeong-dong), Tongyeong-si, Gyeongsangnam-do

📟 N/A 📷 @sammoondang_coffee_company

📅 **Tue to Sun** 11:00am - 9:00pm / **Mon** Closed

Established **2014** Brewing Method **Espresso, Filter** (Kalita Wave)
Recommended Menu **Sammundang Espresso**

Behind the Counter:
Machine **La Marzocco Linea 2 Group**
Grinder **Mahlkönig K30 Twin 2.0, EK43** Roaster **Diedrich IR-5**

Cafes like Sammundang in the rural harbour town of Tongyeong (통영) are a constant reminder that specialty coffee is not just restricted to the larger cities Seoul or Busan. From

Pyeongtaek (Hocus Pocus Roasters호커스포커스로스터스), Daejon (Told A Story톨드어스토리) to Gwangju (Cafe 304카페304) and beyond, it is now possible to enjoy small batch roasted coffee almost anywhere in Korea.

Seven years after establishing the first "local" small batch roastery in the region, Coffee Roasters Sooda, the team recently consolidated their business into a two storey venue under the name Sammundang Coffee Company. Complete with a roasting lab, rooftop and espresso bar, the retro themed workshop-like space has quickly become a massive hit with visitors to the region, particularly from nearby Busan.

It's conveniently located too; cross the road to see the famous 17th Century Choson era (조선시대) Sebyeong-gwan Hall (통영 세병관), wander down alleyways to the harbourside Joongang market (중앙시장) or take a short hike up to Dongpirang Mural Village (동피랑 벽화마을) for the best

vantage point in town. If you've got time to hang around, stay for one of Sammungdang's indie concerts or pair your flat white with a panini in the brunch restaurant downstairs.

Roaster Roasting in-house on the ground floor of the building, Sammundang offers an ever changing variety of seasonal coffee available for espresso and Kalita Wave hand-drip.

Gwangju and Jeju Island

Gwangju and Jeju Island

① Gwangju
② Jeju-si
③ Hallim-eup, Jeju-si
④ Seogwipo-si

①

Chipyeong-dong

Cafe 304 242p

Kim Daejung Convention Center

③ ④

Cafe Imyeon 254p

Be Brave 252p

-dong

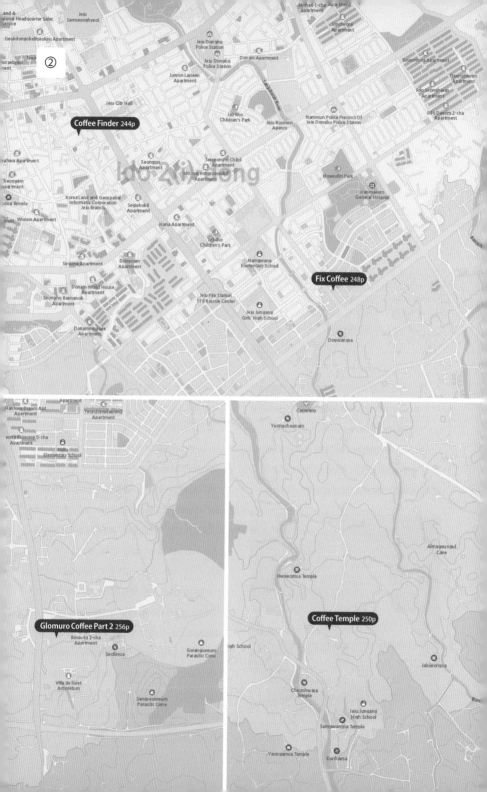

Cafe 304

카페 304

📍 15 Sangmunuri-ro (Mareuk-dong), Seo-gu, Gwangju

🚇 Kim Daejung Convention Center (Line 1, Exit 4)   📷 @304coffeeroasters

🏠 **Mon to Sun** 10:00am – 9:30pm

Established **2010**   Brewing Method **Espresso, Filter (V60), Cold Brew**
Recommended Menu **Today's Coffee (Filter)**

Behind the Counter:
Machine **La Marzocco KB90 3 Group**   Grinder **Mahlkönig EK43**
Roaster **Probat, Trinitas**

A mainstay of Gwangju's coffee culture for over ten years now, 304 Coffee Roasters (hereafter 304) has built a strong presence in the country's Southwest with a brand that now includes five cafes, a roastery, and an in-house bakery.

Well ahead of the curve, the team behind 304 first started importing specialty beans in 2010, trading directly with growers across the world, roasting locally in their lab and creating a loyal base of coffee enthusiasts. Over the years 304 has become the flagship brand in Gwangju's coffee scene, with visitors coming from far and wide to sample their style of delicately roasted single origins. The menu often has a dozen different options including Ethiopia, Costa Rica, and Kenya, as well as seasonal specials like Cup of

Excellence (CoE) Colombia Bella Vista's, Qima Yemen Auction Lots and Panama Geisha.

Located on one of Sangmu-jigu's (상무지구) main thoroughfares, their Mareuk (마륵) branch has ceiling-high French windows facing out onto a bright brickstone patio, ideal for sipping coffee during the long Gwangju summers.

Directions With a population of just over one and a half million, Gwangju's transport network is significantly less complicated than the capital's ten-line subway network. To reach 304 Coffee Roaster's Mareuk branch simply jump on Line 1 (of 1) and head west from the central business district to exit 4 of Kim Daejung Convention Center station (김대중컨벤션센터역).

Other Branches:
- **Cafe 304 (Sinchang - 신창)**
  - 📍 280-14 Jangsin-ro (Sinchang-dong), Gwangsan-gu, Gwangju
  - 🚆 N/A
- **Cafe 304 (Chungjang - 충장)**
  - 📍 12-18 Seoseok-ro 7beon-gil (Bulno-dong), Dong-gu, Gwangju
  - 🚆 Culture Complex (Line 1, Exit 3)

Coffee Finder

커피파인더

📍 20 Seogwang-ro 32-gil (Ido 2-dong), Jeju-si, Jeju-do   📠 N/A

📷 @coffeefinder_korea   📅 **Mon to Sun** 10:00am - 10:00pm

Established **2016**   Brewing Method **Espresso, Filter** (V60, December Dripper)
Recommended Menu **Signature Latte** (Peanut)

Behind the Counter:
Machine **Synesso MVP Hydra 3 Group**
Grinder **Anfim SP II, Mazzer Robur, Mahlkönig K80, Peak**
Roaster **Probat, Stronghold S7**

Opened in 2016, Coffee Finder is one of Jeju's first-generation specialty coffee brands. Spread across two floors in a trendy re-modelled villa, the cafe includes an on-site roastery, bakery and an impressive ground floor espresso and hand-drip brewing counter.

Owner and Jeju-local Ji Joon-ho (지준호) regularly opens the cafe to the local community, holding concerts, art exhibitions and even K-pop fan meetings in the open-plan two-storey space. Having trained in Seoul since 2009, Ji invites partner cafes Coffee Me Up (커피미업) , Ryan's Coffee (라이언스커피) and Unstatic Place (언스태틱 플레이스) for cupping sessions and pop-up barista events on the island too.

Coffee Finder prides itself on a strong line-up of directly sourced green

beans coming straight from award-winning farmers in Costa Rica, Panama and Guatemala. Ask for the hand-drip brewing menu and you are bound to see more than a few Cup of Excellence (CoE) specialty pour overs. For the aficionado, take a look at the tasting notes of these a-grade beans and you'll also find rare processing methods including carbonic maceration, cold press natural and red honey fermentation.

Other drinks menus include three different signature lattes (peanut, sesame and vanilla), around ten fresh fruit juices and smoothies as well as hot chocolates and herbal teas. Thanks to an in-house patisserie, added during the cafe's expansion in 2019, they also offer an array of baked goods too. But first, you have to find Coffee Finder! Two streets back from Jeju city's Jungang-ro (1131 지방도), the cafe is accessed via a small concrete laneway branching off Seogwang-ro 32-gil (서광로32길). If you are taking a taxi or opt for public transport, aim for the Jeju-si City Hall (제주 시청) and you will find the cafe after a short two minute stroll.

Fix Coffee

픽스 커피

2 Sinseol-ro 11-gil (Ara 2-dong), Jeju-si, Jeju-do

N/A @fixcoffee_

Wed to Mon 8:00am - 6:00pm / **Tue** Closed

Established **2016**   Brewing Method **Espresso, Filter (V60), Cold Brew**
Recommended Menu **Fix Coffee / Latte / Espresso**

Behind the Counter:

Machine **Kees van der Westen Mirage 3 Group**
Grinder **Anfim SP II, Mazzer Robur, Mahlkönig K80**

Run by husband and wife duo Jeong Tae-seung (정태승) and An Hee-jin (안희진), Fix Coffee has become somewhat of an institution on the island of Jeju since opening in 2016. One of the first specialty coffee shops to land on the island, the brand recently expanded with a second venue in nearby Hwabuk i-dong (화북이동), cementing their place as a key player in the local coffee scene.

Formerly an engineer, and with years of previous experience in the Seoul coffee industry, owner and barista Jeong runs a workshop too, modifying and repairing espresso machines for a growing number of partner cafes across the island.

Maintaining strong links with Seoul brands, Fix Coffee uses beans from market pioneer Coffee Libre (커피 리브레) for espresso blends, Namusairo (나무

사이로) and Low Key (로우키) for single origin hand drip brews, and a tea selection curated by Hongdae-based altdif (알디프).

Their larger mezzanine style branch in Gongdan regularly hosts pop-up shops, contemporary art exhibitions and coffee seminars. Keep a tab on their Instagram page for more information on dates and ticket availability.

Other Branches:
- **Fix Coffee (Gongdan - 공단)**
  📍 55 Cheongpungnam 8-gil (Hwabuk i-dong), Jeju-si, Jeju-do
  🚉 N/A

Coffee Temple*

커피템플

📍 Joongsun Nongwon Cultural Space, 269 Yeongpyeong-gil (Wolpyeong-dong), Jeju-si, Jeju-do
📞 N/A　　📷 @coffeetemple_jeju
🏪 **Wed to Mon** 10:00am - 7:00pm / **Tue** Closed

Established **2009**　　Brewing Method **Espresso, Filter (V60)**
Recommended Menu **Tangerine Cappuccino**

Behind the Counter:
Machine **Dalla Corte XT 3 Group**
Grinder **Mazzer Robur, Anfim SP II**

Long before opening his first cafe in Sangam-dong in 2009, Kim Sa-hong (김사홍) competed regularly on the domestic and international barista competition circuit, winning both the 2007 and 2008 Korea Barista Championship (KBC) in succession. After winning again in 2015 he was selected to represent South Korea in the 2016 World Barista Championship (Dublin), cementing his reputation as one of the most successful baristas in the country.

After a decade of business in Seoul, Kim eventually closed the original venue and moved the business to

Jeju island in 2019. Surrounded by rolling hills and a hectare of tangerine trees, this peaceful coffee retreat is located inside contemporary art and exhibition space Joongsun Nongwon Farm (중선농원). Sit back and relax with some of the best coffee on the island, including the famous "Super Clean Espresso" (first served up at the 2016 World Barista Championship) and Tangerine Cappuccino. Originally developed at Coffee Temple's first branch in Seoul in 2009, this signature latte is a must-try menu when visiting Jeju. Combining the fruity acidity of a perfectly extracted coffee, a fresh Jeju tangerine, and a small dose of citron syrup, this creative menu is the ideal introduction to Kim Sa-hong's (김사홍) multi-award winning brand.

Guest Barista Events Throughout 2020 and 2021, Kim barista travelled extensively throughout Korea hosting pop-up collaboration "Guest Barista" events (shortened to 게바 in Korean). Working closely with roasters at each of the venues, Kim demonstrated his ability as one of the finest espresso baristas in the country by creating a different signature menu for each of the 18 cafes throughout his 14 months epic tour. Word of mouth quickly spread and soon hundreds of people joined the queue each weekend to sample one-off menus in a festival-like atmosphere. With more events planned for 2022, make sure to check out their Instagram page for more information.

Be Brave

비브레이브

📍 85-13 Seohojungang-ro (Seoho-dong), Seogwipo-si, Jeju-do   🚊 N/A

📷 @be_brave_korea   🕐 **Mon to Sun** 9:00am - 10:00pm

Established **2017**   Brewing Method **Espresso, Filter (V60)**

Recommended Menu **Single Origin Filter Coffee / "Roche" (로쉐) Vanilla bean Latte**

Behind the Counter:

Machine   Victoria Arduino VA388 Black Eagle 3 Group

Grinder   Mazzer Kold S, Mahlkönig EK43

Roaster   Probat, Stronghold, Trinitas T2

This is one for all the filter coffee lovers. Famous for its high quality V60 coffee lineup, Be Brave serves up everything from rare Panama Geisha, Cup of Excellence (CoE) auction lots and Specialty Coffee Association (SCA) 90 + scoring coffees from renowned Panama farms Finca Deborah and Don Pepe. With over ten different single origins available, priced between 8,000 Won and 15,000 Won a cup, take your time to browse the menu or pick

from the colour-coded tasting notes at the counter.

For espresso-based drinks, try single origin americanos pulled by champion barista Kim Hyeon-min (김현민), winner of the 2019 Espresso Throwdown Korea. Top this off with their Victoria Arduino Black Eagle espresso machine, the same model used at the World Barista Championships (WBC), and you have the ultimate recipe for coffee satisfaction.

Five years since opening their first branch in Seoho-dong (서호동) this popular brand now boasts a total of four venues, two of which opened in Jeju's Seogwipo (서귀포), one in Busan (부산 서면) and one in Cheonan (천안 백석), a short drive south of the capital.

Cafe Imyeon

카페이면

📍 13 Geumneung 5-gil (Geumneung-ri), Hallim-eup, Jeju-si, Jeju-do   📓 N/A

📷 @cafe_the_other_side   🗓 **Mon to Fri** 9:00am - 6:00pm / **Sat to Sun** Closed

Established 2020   Brewing Method **Filter (Chemex)**
Recommended Menu **Single Origin Filter Coffee**

Behind the Counter:
Machine Decent Espresso DE1 Pro 1 Group
Grinder Mazzer Robur, Victoria Arduino Mythos One
Roaster Burning Roaster

One of few places in Jeju that specialise in Chemex filter coffee, Imyeon ("The Other Side" or 이면 in Korean) is located in the quiet west coast village of Geumneung (금능), just a short walk from the beautiful Geumneung Beach (금능해수욕장). Converted from a local farm building, Imyeon is popular for its relaxing "tiny house" vibe, peaceful surroundings and coffee pairing desserts. On entering the cafe you can choose from a range of four different single origin coffees displayed in the centre of the room. Each of the small batch specialty beans are roasted in-house and selected carefully from around the world including award-winning farms in Papua New Guinea, Honduras, Ethiopia and Brazil. All their coffee is roasted using a Korean-manufactured

"Burning" Origin small batch roaster tucked away in the rear of the tiled kitchen.

As well as brewing coffee, the team also serves up milk tea, latte, and espresso. Whatever you order, be sure to leave room for one of their crowd-pleasing baked goods including chocolate pound cake, petite madeleine and financiers.

Interior Hidden inside the quaint farmhouse building, the interior features a careful balance of European, Japanese and Korean finishing touches.

Espresso Machine Another unique aspect to Imyeon is their state-of-the-art DE1 Pro espresso machine by Decent Espresso (USA). One of the newest espresso machines on the market, scaled down iPad style machine is backed by one of the world's leading coffee authorities, the one and only Scott Rao.

Glomuro Coffee Part 2

그러므로 Part 2

📍 16-14 Sumogwon-gil (Nohyeong-dong), Jeju-si, Jeju-do   🚆 N/A

📷 @glomuro_coffee   🕐 **Tue to Sun** 10:30am - 9:00pm / **Mon** Closed

Established **2014**   Brewing Method **Espresso, Filter (V60)**

Recommended Menu **Merry HaHa / Seasonal Americano**

Behind the Counter:

Machine   **La Marzocco Linea 3 Group**

Grinder   **Victoria Arduino Mythos One, Mahlkönig EK43**

Roaster   **Probat**

Located on the road south to Jeju's symbolic Hallasan Mountain (한라산), Glomuro's second venture has elevated this brand from local Gunam-dong (구남동) roastery to its current status as one of the hottest cafes in the region.

Disguised inside a quirky, ultra modern brick bungalow, the cafe features a central coffee bar, in-house coffee roasting lab and open-plan seating area. A minimalist theme continues throughout with monotone bleached walls and floor-to-ceiling windows that flood the space with natural lighting and offer views of the terrace and surrounding garden area. Alongside the standard menu, Glomuro serves up seven different "Signature" menus including their

number one bestseller, the "Merry HaHa". A combination of condensed milk, syrup and house blend espresso, this is a unique twist on the caramel macchiato, served cool and in a standard 6 oz coffee cup. When it comes to the classics, we recommend ordering their americano. Taking freshly roasted single origin beans, the team of baristas pull clean and balanced espresso shots from their La Marzocco Linea that are bound to satisfy your coffee cravings. Lattes, cappuccinos and flat whites are perfect too. To pair, don't miss out on their famous selection of desserts including fresh fruit tarts, almond cookies, and the delicious caramel mousse cake.

If you fancy a stroll to walk it all off, head towards the picturesque Halla Arboretum (한라 수목원), just a few hundred metres to the east of the cafe.

Other Branches:
• **Glomuro Coffee**
  📍 45 Gunam-dong 6-gil (Ido 2-dong), Jeju-si, Jeju-do
  📖 N/A

KOREA SPECIALTY COFFEE GUIDE

초판 1쇄 발행 | 2022년 10월 5일
초판 2쇄 발행 | 2022년 11월 15일

지은이 | 찰스 코스텔로, 조원진, 심재범

펴낸곳 | 도서출판 따비
펴낸이 | 박성경
편 집 | 신수진, 정우진
디자인 | 이수정

출판등록 | 2009년 5월 4일 제2010-000256호
주소 | 서울시 마포구 월드컵로28길 6(성산동, 3층)
전화 | 02-326-3897
팩스 | 02-6919-1277
메일 | tabibooks@hotmail.com
인쇄·제본 | 영신사

ISBN 979-11-92169-19-4 03590
값 25,000원

* 이 책에 실린 카페 지도는 국토지리정보원에서 제공받은 지도를 바탕으로 제작하였습니다.
* 테라로사(강릉), 베르크로스터스, 빈브라더스, 센터커피, 카페 304, 톨드어스토리의 사진은
  각 업체가 제공하였고, 나머지 카페들의 사진은 저자들이 찍었습니다.